CW01192847

HEYDAY OF THE SHOTGUN

HEYDAY OF THE SHOTGUN

The art of the gunmaker at the turn of the last century

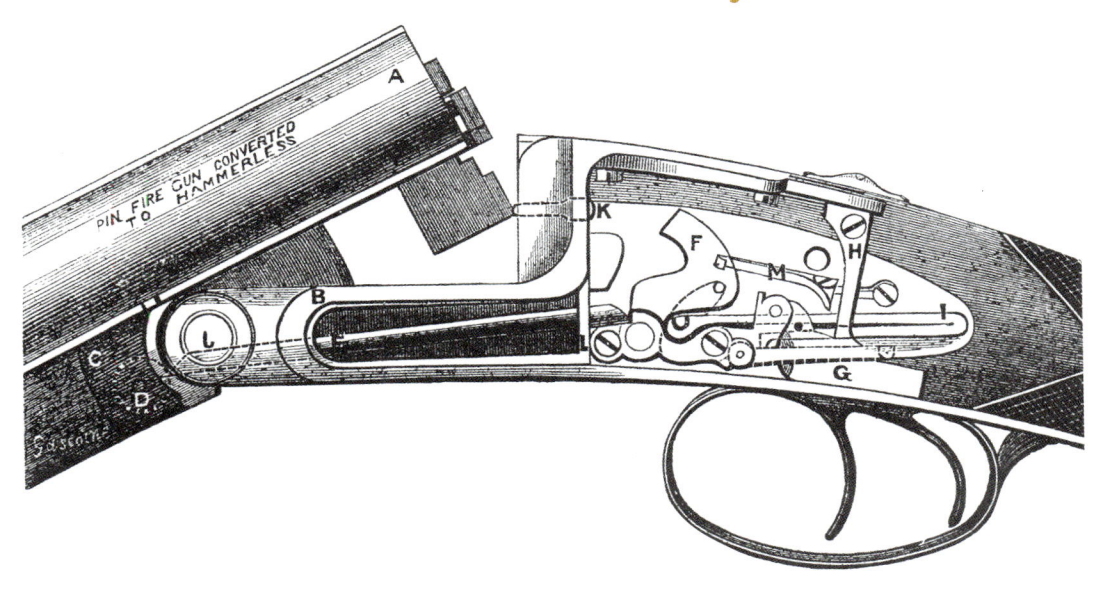

DAVID J. BAKER

SWAN·HILL PRESS

Copyright © 2000 David J. Baker

First published in the UK in 2000
by Swan Hill Press, an imprint of Airlife Publishing Ltd

British Library Cataloguing-in-Publication Data
A catalogue record for this book
is available from the British Library

ISBN 1 85310 974 6

The information in this book is true and complete to the best of our knowledge. All recommendations are made without any guarantee on the part of the Publisher, who also disclaims any liability incurred in connection with the use of this data or specific details.

All rights reserved. No part of this book may be reproduced or transmitted in any form or by any means, electronic or mechanical including photocopying, recording or by any information storage and retrieval system, without permission from the Publisher in writing.

Typeset by Servis Filmsetting Ltd, Manchester.
Printed in China.

Swan Hill Press
an imprint of Airlife Publishing Ltd
101 Longden Road, Shrewsbury, SY3 9EB, England
E-mail: airlife@airlifebooks.com
Website: www.airlifebooks.com

Acknowledgements

This book is the result of a thirty-year interest in the evolution of the sporting gun, therefore it could well be claimed that any list of acknowledgements is almost certainly incomplete.

To all the good friends made over the years, I would tender my sincere thanks for all their help contributed in all sorts of ways. The following list is deliberately laid out in random order:

The staff of Cambridge University Library	Paul Mosse
The staff of The Musée d'Armes de Liege	Mrs A. Thorn
Sarah Gibbins of The Clay Pigeon Shooting Association Ltd	Emily Baker
	Adrian Lemmon
Angela Kelsall of The Derby Museum and Art Gallery	George Yannagas
	Mark Wright
Christina La Torre of The National Army Museum	Phil Gay
Melanie Baldwin of The Yorkshire Museum	Dave Harnett
	Neil Beeby
Mark Crudgington	Huw Williams
Ian Crudgington	Professor Dr Martin Moog
Rob Knowles	Nigel Brown
Rodney Ford	Frank Page
Mike Delanoy	Alec Morris
Jim Buchanan	John Wiseman
Peter McGowan	Hugh Barrell
John Dillon	

Foreword

The heyday of the British shotgun appears to have passed. This book celebrates the era, which saw a wonderful diversity of guns, designed, developed and produced in Great Britain.

Two world wars and changes that moved faster than the gun trade could keep up with, saw the near cessation of gun production in this country. We influenced production abroad but did little to save the home-grown product.

Thankfully, the 1990s have seen the chinks of light needed to spur on production. Best guns are now produced by a large number of makers, not just those based in London. New designs have gone into production and old ideas have been improved upon. It seems that all we need is a small revolution, as witnessed with the introduction of the breech-loading gun at the Great Exhibition to launch the inventive mind of the trade forward again. Maybe we can experience another age of the British shotgun in the 21st century. I fervently hope so.

This book is a poignant reminder of the British shotgun's illustrious past and the possibilities for the future.

Mark Crudgington

Contents

Introduction		8
CHAPTER ONE:	The Origins	9
CHAPTER TWO:	A Gun for a Nobleman	16
CHAPTER THREE:	A Gun for a Gentleman	25
CHAPTER FOUR:	A Gun for a Young Officer	36
CHAPTER FIVE:	A Gun for a Gamekeeper	46
CHAPTER SIX:	A Gun for a Poacher	55
CHAPTER SEVEN:	A Gun for a Conservative	62
CHAPTER EIGHT:	A Gun for a Trap Shooter	73
CHAPTER NINE:	A Gun for a Wildfowler	82
CHAPTER TEN:	The Cartridge of the Heyday	94
CHAPTER ELEVEN:	The Gun for the Future	106
Epilogue		118
APPENDIX ONE:	A Gunmaker's Catalogue	120
APPENDIX TWO:	The Gunmakers of the United Kingdom in 1900	148
Bibliography		153
Index		154

Introduction

The heyday of the shotgun

About a decade before the first shots were fired in anger on that fateful day in August 1914, the evolution of the sporting gun and the society that created and sustained it in Great Britain attained a peak that was soon destined to be changed forever by the mud of Flanders. This heyday was not a single day, month, or year. It was a period, just as the twentieth century was born, when everything that had gone before in the gun trade over the preceding fifty or more years came to fruition.

This book is an attempt to take a snapshot of this world from the point of view of the British shotgunner, whatever the individual's social station. It makes no claim to be a definitive account because, after the lapse of nearly a century, it simply could not be so. Instead, it is an overview compiled from all the contemporary sources that can be mustered. Inevitably, it is coloured by vested interests, sometimes because the sources were produced as advertising copy, and yet more distortion is introduced by my personal opinions. While I have attempted to restrain these aberrations, I have not sought to eliminate them, in the hope that they add character.

What I have attempted is to focus on a set of values so far removed from those we know, as to be almost alien. The hope is that I can conjure up enough of this unfamiliar world to give the reader an awareness of it, thereby furthering your understanding of the sporting gun that was so much part of it.

CHAPTER ONE
The Origins

The reason the British shotgun enjoyed a heyday as the nineteenth century grew into the twentieth was because the conditions were right. Unlike the instant flowering which blossoms after a brief desert rain, the perfected gun evolved after a long gestation period encompassing the preceding two and a half centuries. Moreover, this peak of development was not an isolated phenomenon. It was made possible by and was part of the broad advance of technology that made the nineteenth century different to those that had gone before. To cite just one example, it was no accident that man's age-old dream of powered flight became a reality in the same period.

The shotgun evolved to the perfection it attained as the twentieth century was born not simply because the technology existed to create it, but because the financial incentive was there to harness the potential. In short, the social conditions were such that fortunes could be made by successful gunmakers and inventors of all sorts.

Of all the men and women who played a part in the evolution of the British shotgun, perhaps the most important was HRH Prince Albert. When this German prince came to England in 1840 to marry Queen Victoria, he brought with him a love of field sports, and his guns and rifles. However, the excellence of the London gunmakers ensured that, very soon after he arrived, Albert was commissioning work from them. In due time the 'Prince Consort', as he came to be known officially, was to pass on to his sons a love of the chase and

Pheasant shooting with ladies watching. Sidney Smith. (PHOTOGRAPH BY COURTESY OF THE YORKSHIRE MUSEUM)

they gave the vital royal leadership to the social order that sustained the gun trade.

This was not Albert's only contribution. As part of the renaissance surrounding the Great Exhibition of 1851 in which he played a crucial role, there was a reorganisation of the British patent system. A labyrinth of daunting expense and complexity was replaced by a new set of regulations designed with the express purpose of stimulating invention. The revolutionary feature of The Patent Act of 1852 was that patents should be cheap to obtain and only become expensive for the inventor towards the end of the period of protection. Since the protection could be allowed to lapse at many earlier points in the process, the incentive existed to protect and thereby publish any idea that seemed to have potential. The system was not confined to British subjects, anyone in the world could obtain a British patent, thereby enriching the domestic pot. Inevitably, of course, the bulk of what was protected proved to be of limited commercial use at best. Moreover, the cachet 'Patent', or more grandly 'By Royal Letters Patent', became a favourite advertising claim, and in many cases the patent's greatest value. But none of this detracts from the brilliant concept of the scheme, which was to create an accessible corps of knowledge which permitted inventors to profit from their inventions, and gave stimulus to others to improve still further.

The two new stimuli to the evolution of the sporting gun, the royalty-led social scene in which sport played an important role and the encouragement of invention, were, of course, but additions to the existing order. For some 200 years wild game had been bagged for the pot with a shotgun by those permitted by law to carry one. Travel was limited, so social shooting was confined to those who lived within riding distance of each other. However, enough men shot to support a healthy gun-making industry, and there were sufficient rich men to encourage gunmakers to improve their wares to the state where, at the beginning of the nineteenth century, the fine sporting guns retailed in London were reckoned to be the best in the world. These guns were, of course, muzzle loaders, fired by flint striking on steel and a goodly proportion of them were single barrels.

In the early 1800s, the application of the growing science of chemistry had paved the way for another age-old dream of inventors: the gun

The perfected Joseph Manton double-barrel flintlock gun. This 20-bore example built in 1813 incorporates Manton's elevated rib and gravitating stops. These have platinum-filled weights which, when the gun is vertical, engage catches in the breasts of the cocks, thereby locking them and preventing accidental discharge as the gun is reloaded. (PHOTOGRAPH COURTESY OF MESSERS BONHAMS)

THE ORIGINS

G. Bales of Ipswich, *c.* 1840. A 14-bore double-barrel percussion ignition muzzle-loading shotgun

Detail of the underside of the Bales gun

that fired more rapidly because it could be loaded at the breech. It had long been known that there were chemical substances that exploded when they were struck with a sharp blow. However, until the early years of the nineteenth century, all attempts to find any practical applications of this property had failed, and the only use it was put to was in that area of chemical demonstration that is akin to conjuring. When, at last, it was discovered how to tame and make use of these compounds to fire a charge of common gunpowder, it was realised almost immediately that this opened the way to the production of a practical breech-loader. However, a combination of conservatism on the

S. Ebrall of Shrewsbury 12-bore pinfire, *c.* 1865. Bar-action locks and screw grip, Henry Jones under-lever action

part of the shooting public and technical difficulties thwarted this rapid development.

Instead, the shotgun went through an intermediate stage in which the ancient method of loading via the muzzle was retained in conjunction with the new and more rapid means of ignition. The new guns were called 'detonators', a term which has now come to mean the ignition, not of a propellant charge, but one which is a high explosive. The muzzle-loader with percussion ignition, to use later terminology, was to remain the normal sporting gun until the middle of the century.

By then the French gunmakers had done more work on the technicalities of producing a sound breech-loading gun and, towards the end of the 1850s, there was an upsurge of interest in the new style of gun. At first, French guns were imported, but very soon they were copied and refined by the British trade. The process of evolution took full advantage of the new patent system, and in the second half of the nineteenth century, there were some 700 patents devoted to the sporting gun. It is both a measure of the rate of change and a comment on the commercial value of many of the inventions that most patents were allowed to lapse early in their life. Nevertheless, they demonstrate the energy with which the sporting gun was being developed.

If we look at the social scene, the popularity of game shooting explains this activity on the part of the gun trade. Prince Albert had died tragically young in 1861 and the Queen denied her heir, Prince Albert Edward ('Bertie'), any meaningful role in the affairs of state. Field sports came to absorb a good proportion of his energies and thus we see the rise of the cult of the shooting party. The development of the railways played a crucial role here. By the second half of the century, Great Britain had an extensive and efficient railway network, so journeys that had once taken days now took hours. Moreover, the comfort of the

THE ORIGINS

T. Murcott of London 12-bore hammerless gun with bar action locks, Murcott's patent (1871) action. Laminated steel barrels

T. Murcott's trade label

railway when compared with horse-drawn road travel was a revelation. The proximity to a station became a desirable feature of a country house. 'So many hours from town' was stressed in agents' advertisements for establishments for sale or to rent. As a result, sportsmen from the Prince downwards could travel with ease to the moors of Scotland or the stubbles of East Anglia.

One further innovation completes the picture, and this was the introduction of driven shooting. Back in the eighteenth century, shooting had been undertaken by a small group of men with dogs, and perhaps a gamekeeper or two in attendance, who walked the fields, moors or woods in search of their quarry. In the first half of the nineteenth century, it had been found possible to drive game so that the birds flew towards the shooters. The techniques required to manage the shoot reliably in this way were well developed by the middle of the second half of the century, and became the normal way to conduct a smart shoot.

Three very different outcomes resulted from this change, but all had a bearing on the evolution of the gun. Properly managed, driven game presents the shooter with more testing marks than the old walked-up birds and a premium is placed on the rapidity with which the gun can be fired. Finally, because the game is coming to the gun, it is possible to convey the shooter from one firing point to the next. As a result, shooting ceased to be the preserve of the active man. Now a virtual invalid could take part. In addition, onlookers, even ladies in their Victorian finery, could be ferried from stand to stand.

These developments influenced the way the gun trade worked. Since nobody likes to appear a duffer in public, the shooting school arose as a valuable adjunct to the business of actually making guns. Indeed, independent schools appeared, whose business was solely to teach shooting.

Charles Lancaster's shooting school. Photograph originally published in *Commerce*, 1896, and subsequently in *Great Guns*. (PHOTOGRAPH BY COURTESY OF GEORGE YANNAGAS)

However, most significant of all in terms of the evolution of the gun was the increased firepower. Indeed, it could be said that the story of the evolution of the sporting gun through the whole of the nineteenth century is nothing more than a pursuit of ever greater firepower. The most obvious manifestations of this were firstly the introduction of the double-barrelled gun, then the flirtation of the trade with guns with more than two barrels and, lastly, the attempts to interest the British sportsmen in magazine guns. More successful was the selling of guns as matched pairs or even trios, so that firepower could be achieved by having a servant to reload. No matter who was reloading, the addition of a mechanism to fling out the spent cartridge was another obvious move towards increased firepower, as indeed had been the adoption of breech-loading itself.

All of these changes were good for business for the gunmaker and it spurred on those with talent, and frankly some of those without, to increase the flood of invention. The overall effect was a classic example of refinement by competition which, by the end of the nineteenth century, led to sporting guns that combined mechanical efficiency with strength and elegance in a way few other tools developed by man have even approached.

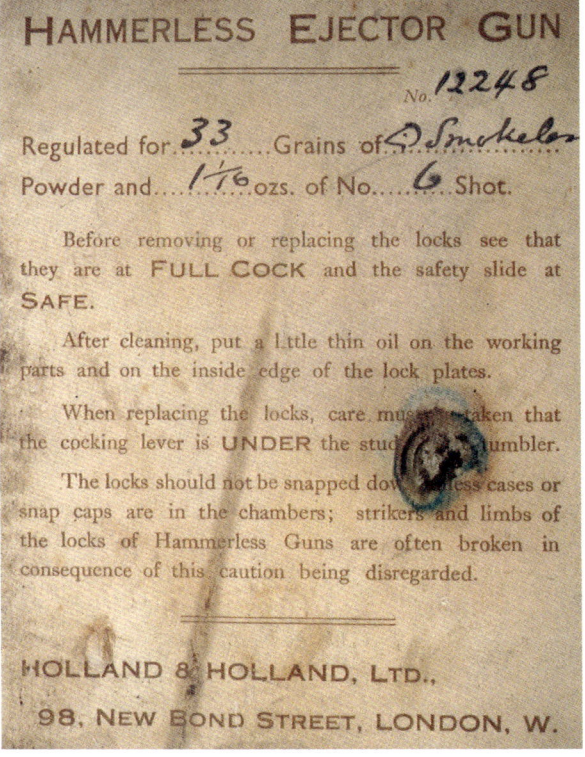

Holland & Holland instruction label from the case of the 'Royal' hammerless gun

Holland & Holland of London *c.* 1890 12-bore 'Royal' hammerless gun

CHAPTER TWO
A Gun for a Nobleman

There is more than a little truth in the notion that a sporting gun is an atavistic link to man's earliest times. In the perilous world into which man emerged, the very best weapon could make the difference between hunger and plenty, or life and death. It is a fact that all down the ages, hunters have aspired to the best available at the time. For the shooting man in Great Britain in 1900 the choice was magnificent. London, at the centre of a vast empire, had a profusion of gunmakers, moreover, the quality of their wares was matched only in Edinburgh, and there on a smaller scale.

Situations like this do not just happen by accident, they emerge and evolve over a period of time. The gun trade was not the only focus of excellence in the capital and all the best of everything from furniture, carriages, silverware to footwear, was made in London. Good parallels can be drawn between the gun trade and the bespoke tailoring trade. In a suit made by a top-flight tailor, there is the combination of the finest materials, expert craftsmanship and the special ingredient of 'cut' or 'line', which transforms what is essentially a person protected from the elements into an elegant man.

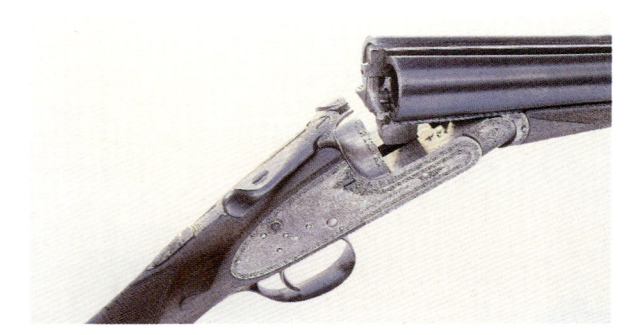

Right: Joseph Lang sidelock hammerless ejector-one trigger

Below: Exchanging guns. Sidney Smith.
(PHOTOGRAPH BY COURTESY OF THE YORKSHIRE MUSEUM)

A GUN FOR A NOBLEMAN

Bell & Prichard and W.R. Pape's advertisement from _The Shooting Times_, 1898

Boss & Co. advertisement from _The Shooting Times_, 1898

Many attempts have been made to try to describe the essence of the best British gun, and, while its beauty is instantly recognisable, translating these qualities into words has taxed all who have tried. It was, it is believed, the eminent Major Sir Gerald Burrard, author of the classic trilogy _The Modern Shotgun_ and sometime editor of _Game and Gun_, who coined the curious phrase 'the look of a snake' as an attempt to capture in words the lean line of a best gun. The problem in trying to define the magic is the fact that entering into the potion are the tactile delights of handling quality. Another attempt at a definition is 'a work of art which feels lighter than it really is'. This takes account of the fact that the balance of such a gun gives an impression that this inert object comes alive in the user's hands.

In many ways it is easier to define what a best gun is not than to explain what it is. For instance, it is nothing to do with show, neither fine wood with breathtaking grain patterns, nor engravings, inlays or other embellishments. It is not simply a matter of the name on the gun. While famous names may give confidence to a potential purchaser and wood patterns and engraving delight the eye, these features may often be associated with a best gun, but do not define it. Essentially, a best gun is the product of perfectionist craftsmen who are also artists, so the gun can be devoid of engraving, have a stock that has less figure than a floor board, and yet be supremely elegant. Such a gun, devoid of the clutter of embellishment, has nothing to distract the eye from the lines and proportions.

James Purdey 16-bore hammerless ejector. The lockwork is the classic Beesley design of 1881. Note the 'compressors' just above the knuckle joint

We are on safer ground when we turn to the origins of the gun we call 'the Best British', for there is general acceptance of the vital role played by Joseph (Joe) Manton. James Purdey the elder said of Manton, 'But for him we should all have been a parcel of blacksmiths'. Col Peter Hawker called him, 'A king of gunmakers'. (Although he was a Lieutenant-Colonel, he is always referred to simply as 'Col Hawker'.)

When Joe Manton opened his first shop in 1792, in Davies Street, Berkeley Square, good guns had been made in London for some 300 years. Some of these guns displayed great technical skill, for instance, the production, with the tools of the mid-sixteenth century, of a double-barrel wheelock pistol, with both barrels designed to take two loads, one over the other. Joseph Manton's contribution can be likened to that of a man who inherits a sound building and adds to it another storey. The new dimension added by Manton was a passion for beautiful workmanship. Manton's reason for the pursuit of perfection was not some abstract whim, it was inspired by the basic economic fact that if Joe Manton's guns shot better than those of his competitors, there was every reason to hope that Joe Manton's business would flourish. In essence, a flintlock is a simple mechanism, a way of striking a fragment of flintstone against steel. What was discovered was that, by careful design and precise manufacture, the efficiency of the lock could be greatly improved. In other words, there was a direct link between the quality of the lockwork and the size of the bag. Add to this the realisation that this advantage could be enhanced by the qualities we call fit and handling quality, and we glimpse the reason that Joe Manton could boast that, 'I shall continue annually to increase my charges by five guineas, and still no gentleman will be without a Joe Manton'. Such were the rewards that derived from the pursuit of excellence in gunmaking guided by Manton's own first-hand experience in the field and the advice of his talented customer, Col Peter Hawker.

Henry Atkin 12-bore hammerless ejector one trigger. The lockwork again is the classic Beesley of 1881
(Photograph by courtesy of Messrs. Christies)

Frederick Beesley sidelock hammerless gun, stripped to show bar-action lockwork, ejector and one trigger mechanisms

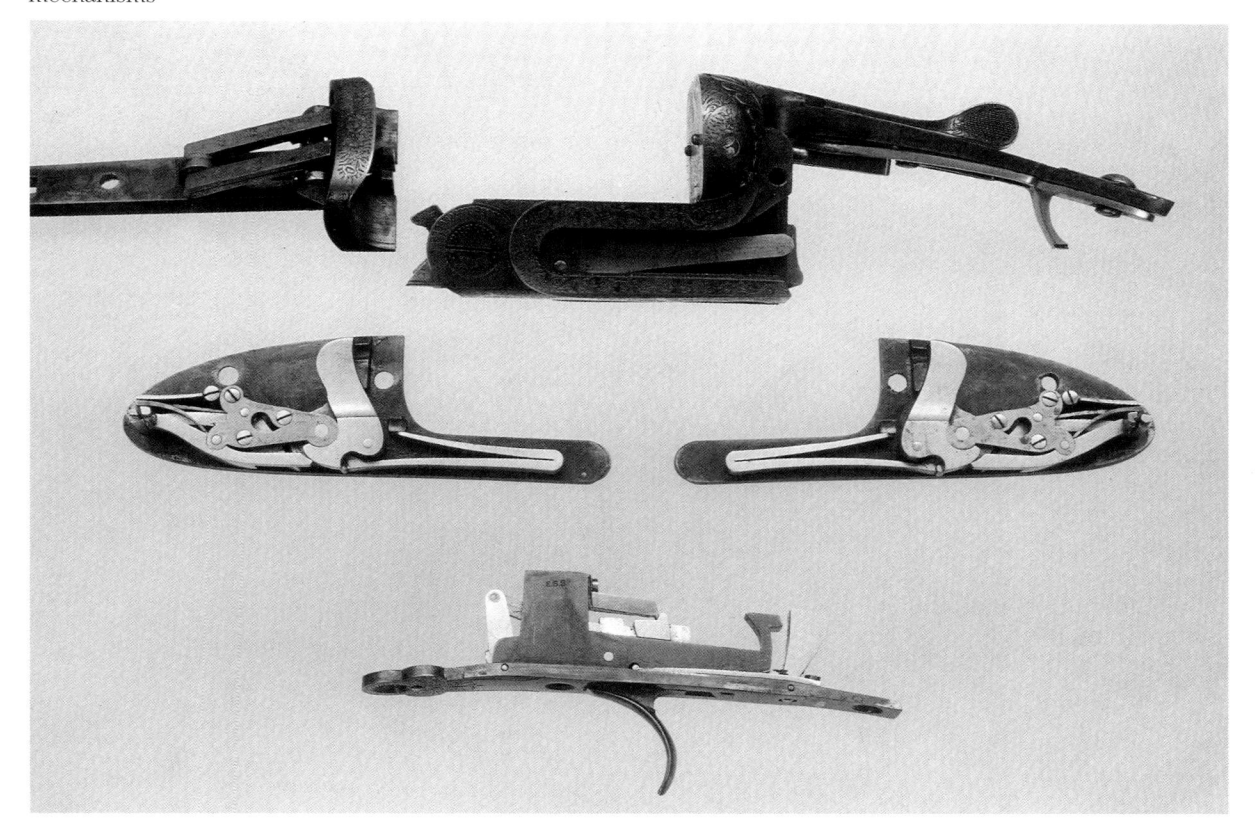

In a few short years everything changed, percussion ignition enabled guns to fire more quickly and, when properly bored, to shoot harder. In Col Hawker's lament for the old order he wrote, 'But nowadays every common fellow in a market town can detonate an old musket, and make it shoot as quick as can be'. He went on, 'We every day meet with Jackanapes-apprentice-boys who can shoot flying, and knock down their eight birds out of ten'. Faced with such facts it would not be unreasonable to expect that shooters would forsake the high-priced best guns. To a degree this happened and the era of percussion was also to see the rise in fortune of the provincial gunmaker. But not every gentleman was content with a 'market-town musket'. In the words of W.A. Adams, who wrote under the pen-name of 'Purple Heather': 'It's the old, old story; if a man's palate has been accustomed to wines of the choicest vintages, he will not drink *vin ordinaire*, or fruity port; and a Burmese cheroot will have no charm for the connoisseur who revels in the delicious perfume of a fine Havannah (sic).'

It would be unfair not to admit that the best provincial work was good, but the fact remains that the best London products had just that extra touch. So, local pride not withstanding, sportsmen who could afford the best of everything continued to buy their shotguns in London, because for a shooter, the use of a fine shotgun is one of the most seductive pleasures of this life. Joseph Manton died in 1835, but a coterie of his workmen went on to found gunmaking firms of their own. Amongst them were Thomas Boss, William Grey, William Moore and James Purdey and they in turn taught their apprentices and inspired both them and their rivals to maintain the demanding standards set by Joe.

With all the inventions of the previous fifty years it might seem strange that in 1900 the shotguns on offer in the gunmakers' shops in the West End of London were all so similar. The

Charles Lancaster back-action sidelock one trigger ejector

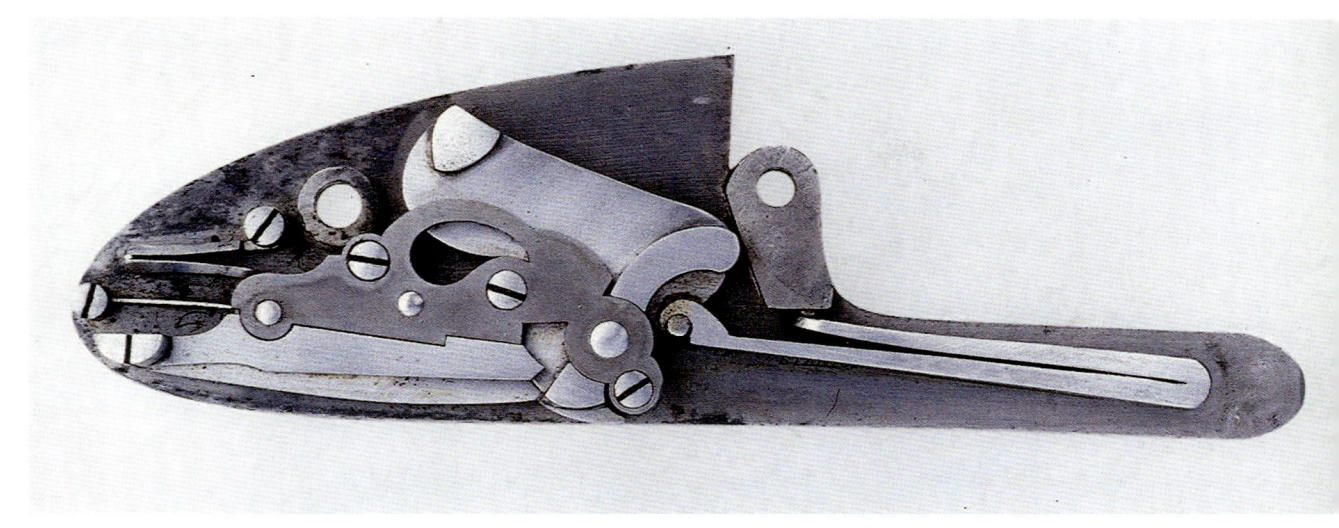

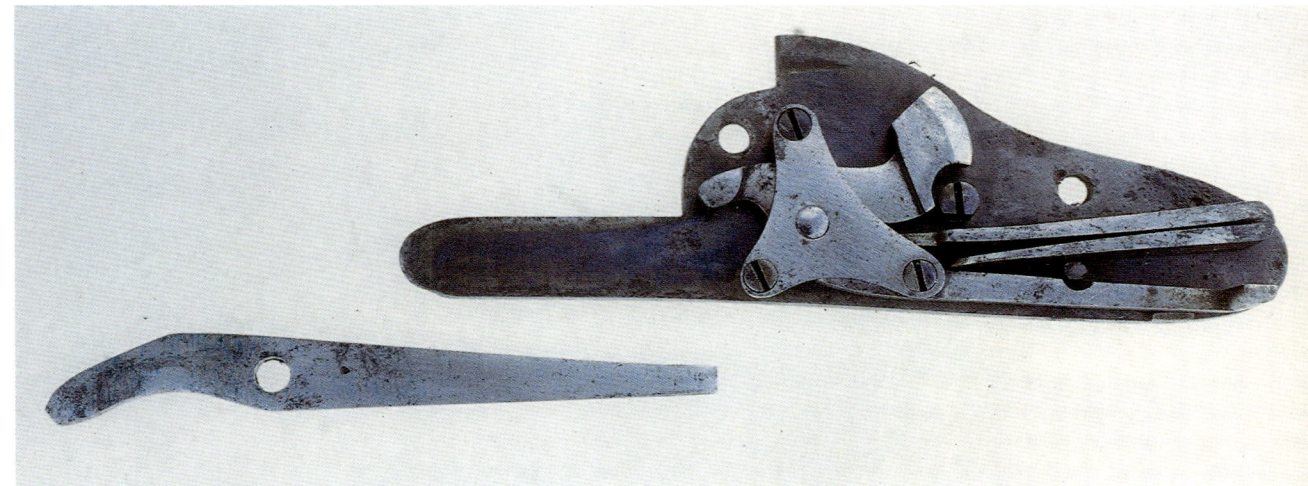

Two sidelocks, essentially similar design but of greatly differing quality and refinement

reason was that, while the technical innovations had altered the mechanical aspects of the gun beyond recognition, it remained, in essence, the double-barrel, side-by-side that Joseph Manton had known. By 1900, the experience of shooters and gunmakers had refined the improvements to best suit the sportsman's purpose. The result, costing between £60 and £70, was the classic top lever opening, side-by-side sidelock hammerless ejector. While superficially alike, the products of the various firms differed in detail. People were still undecided about the wisdom of having one trigger to work both locks. Arguments raged over the rival claims of the several one-trigger mechanisms on sale in the London gunmakers, and the greater number only known in the files of the Patent Office in Southampton Buildings.

In terms of variants, much the same could be said of the ejector. There had been a flood of these in the 1880s and 90s, and indeed there was still a trickle of new designs, but in this case practical experience had decided that a mechanism to fling out the spent cartridge case was a valuable addition to the gun. As noted earlier, the search for firepower runs through the story of the gun and the ejector makes a very significant contribution. Those designs in use had been standardised to the extent that all were simple

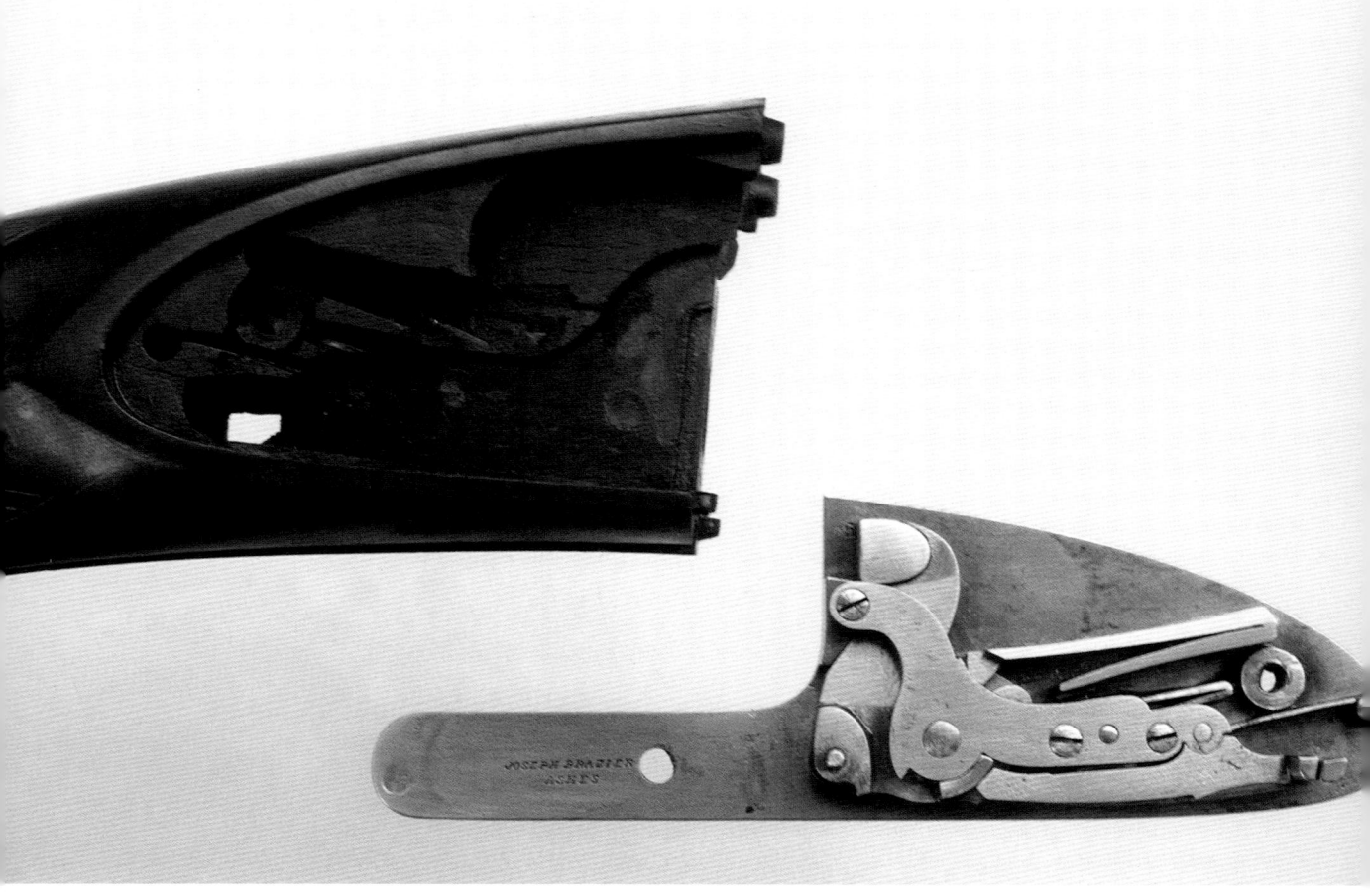

George Gibbs of Bristol action and back-action lockwork, unusual in a best gun

mechanisms mounted in the forend of the gun. Such differences as there were could be dismissed as existing more as talking points in the sale room than for exhibiting any real advantages.

While all the top-flight makers offered sidelock hammerless guns, within this broad classification there were more significant differences. The most distinctive was the Charles Lancaster with its characteristic 'leg-of-mutton' lock plates. By 1900, this look was somewhat dated, having originated in the mid-1880s, but there had been a continual process of refinement to the Lancaster and so the internal mechanism was much more modern than the appearance of the gun suggested.

By contrast, the Purdey gun was an older design than the original Lancaster but, because of the more conventional bar-action lock plates, it looked more up to date. Despite their apparent differences, the stories of the Lancaster and Purdey guns were curiously intertwined, both with each other and with the greater story of the London gun trade. Both of the original designs were conceived by Frederick Beesley, who served his apprenticeship with Messrs Moore and Grey, both ex-Joe Manton men. Then, as part of the refinement of the Lancaster design, a mechanism akin to the Purdey was used – the original Purdey patent having expired. So the differences were somewhat blurred, both the Purdey and the later Lancaster used lock mechanisms in which a cunningly-designed mainspring fired the lockwork, served as the lever to cock it and assisted in the opening of the gun.

The rest of the London trade offered variants on

a somewhat simpler system, in which the mainspring merely powered the lock and a separate lever cocked the mechanism. Even here there is a need to mention the Lancaster, because the final evolution of this gun used a variation on this theme.

Mention should also be made of the side lever opening gun, very much a matter of personal preference which all of the makers would produce to order. This came to be strongly associated with the firm of Stephen Grant and throughout that firm's history a goodly proportion of the guns were set up in this way.

There are many skills needed to build a shotgun and, as in all trades and professions, specialisation enables a practitioner to hone and develop skills to a level not possible for a jack-of-all-trades. The making of springs, engraving, blacking barrels or the fitting of the woodwork are obvious examples of very diverse talents, but one reason for the quality of work found on a best gun is the level to which specialisation was taken. For instance, a man would do but one style of engraving, another would devote himself to ejector work and yet another to the mysteries of the single trigger. No matter which of the gunmaking skills a man or woman practised in the building of a best gun, the bond shared by all was a passion for perfection. The depth of this fervour can be likened to that of a religious convert and stories abound to illustrate it; men who refused to take a holiday because they were happier at work, and yet others who claim never to have done a good day's work in their lives because, by the standards they set for themselves, they had failed. Lesser mortals would look at their work and see only perfection. Skill with a file was a vital necessity and the ability to 'file flat' was essential. An example of this is to have the ability to take two rough cubes of steel, reduce them to a quarter of their original size and in doing so produce perfect cubes with faces so true that, if one cube were placed on top of the other, both could be lifted up by lifting the top cube. Such skill acquired by a lifetime's application is scarcely credible. The minute sub-divisions of the work of building a fine sporting gun, or for that matter a rifle, expose the utter nonsense that we talk when we describe the name engraved on the item as the 'maker'. Retailer he, or more rarely she, will certainly have been, but very few actually made any contribution to the work of making the gun. There were exceptions to this state of affairs, for instance, Charles Boswell in his little shop above Starkies Chemist in The Strand worked at the bench between dealing with customers.

In fact, the Boswell set-up could be taken as a good example of how the trade was organised, with some work being done on the maker's premises and some being put out to out-workers. The proportion of the work done by these independent specialists varied from maker to maker, the nature of the work involved and the state of the gunmaker's order book.

Historically, the gun quarter of London had been in the district known as The Minories adjacent to the White Tower (Tower of London), where the London proof house is situated. By 1900, however, out-workers to the gun trade were scattered throughout the less fashionable areas of the city. Here they were but part of a wider picture which would have been revealed by a walk down any of the side streets. London was a city of artisans, craftsmen and small factories.

The out-workers to the gun trade were a diverse lot. Some, such as Edwin Hodges, the gun-action filer from Islington, were well known. Hodges, for instance, was the co-patentee of the action that Stephen Grant had used on the bulk of his hammer guns in the 1870s. Indeed, Hodges blurred the line between gunmaker and out-worker because, for a small select clientele, he produced specialist live pigeon trap guns.

Other out-workers such as John Sumner, engraver in Bateman Street, Soho, or Frank Squires, barrel maker in Berwick Street, were known to the greater public because their names and addresses were recorded in the street directories.

However, *Kelly's* told only part of the story. Fleeting references can be found in other sources to out-workers producing work of the best quality who are simply unrecorded. Presumably these men were known to their peers and customers and needed no further advertisement.

One unfortunate outcome of the way in which a best London gun was created was that outside the trade very few knew who had actually made which gun. The rich man who bought a pair of best guns in the West End would probably have less idea of where and how the beautiful locks on his new guns were made than he did about the source of the coal that burnt on the fire in the gun room. If he were to take the locks off his acquisition, he would have perhaps noted the word 'Chilton' neatly stamped inside. It is very doubtful if he would have known that these locks were one of the few finished parts of his gun that had not been made in London, or that Chilton, along with Brazier and Stanton, was not from Birmingham but Wolverhampton, where, for the preceding century, lock making had been a specialist industry. Few other parts of the gun would be signed, typically there would be a stamp with the initials of the barrel filer hidden under the forend. Beyond this there might well be no other visible advertisement of the craftsmen involved. Most surprising of all, given the obvious parallels between this and the world of fine art, even special engraving was unsigned.

There was obviously a balance to be struck. A craftsman might wish to work undisturbed and acknowledgement of his craft could have encouraged a string of time-wasting visitors. The master gunmaker would have had a good reason for concealing the identity of a stocker because there would have been the danger of the customer taking his repair work to the craftsman direct. Many owners of a fine gun would no doubt have valued the chance to have even a brief insight into how it was made.

From *The Shooting Times* of 1898. Note that the Emperor is shooting with one hand

A page from A. Nobel advertising booklet. *c.* 1905

CHAPTER THREE
A Gun for a Gentleman

At the turn of the century, a suitably slim volume entitled *Shooting on a Small Income* was published by Archibald Constable & Co. and was unusually well advertised in the sporting press of the day. Of course, 'small' is a relative term, the income envisaged by the author, C.E. Walker, was sufficient to employ a gamekeeper at sixteen shillings (80p) a week and to rent a sufficient acreage of shooting to need his full-time attention.

However, true to his book's title, Mr Walker always had a wary eye on his balance sheet. His advice regarding the choice of gun was that it was unnecessary to lay out more than forty or fifty guineas to be well suited, and foolhardy to pay less than fifteen. Within this price range the recommendation was to go for a hammerless gun, because the author regarded the hammer gun as dangerous, to avoid a single trigger because they were unreliable and to buy an ejector only if much rapid shooting was expected. No individual maker was recommended, the only guidance was that the maker should be well known.

Now, while it is certain that at least some of Mr Walker's contemporaries would not have agreed with his views, they are undeniably sound and, if nothing else, a good starting point for this chapter.

Inevitably, the price of a new gun is composed of many factors and a significant item can be the cost of the maker's premises. Large sites on

Group of guns with beaters. Sidney Smith. (PHOTOGRAPH BY COURTESY OF THE YORKSHIRE MUSEUM)

For 1904 GUNS.

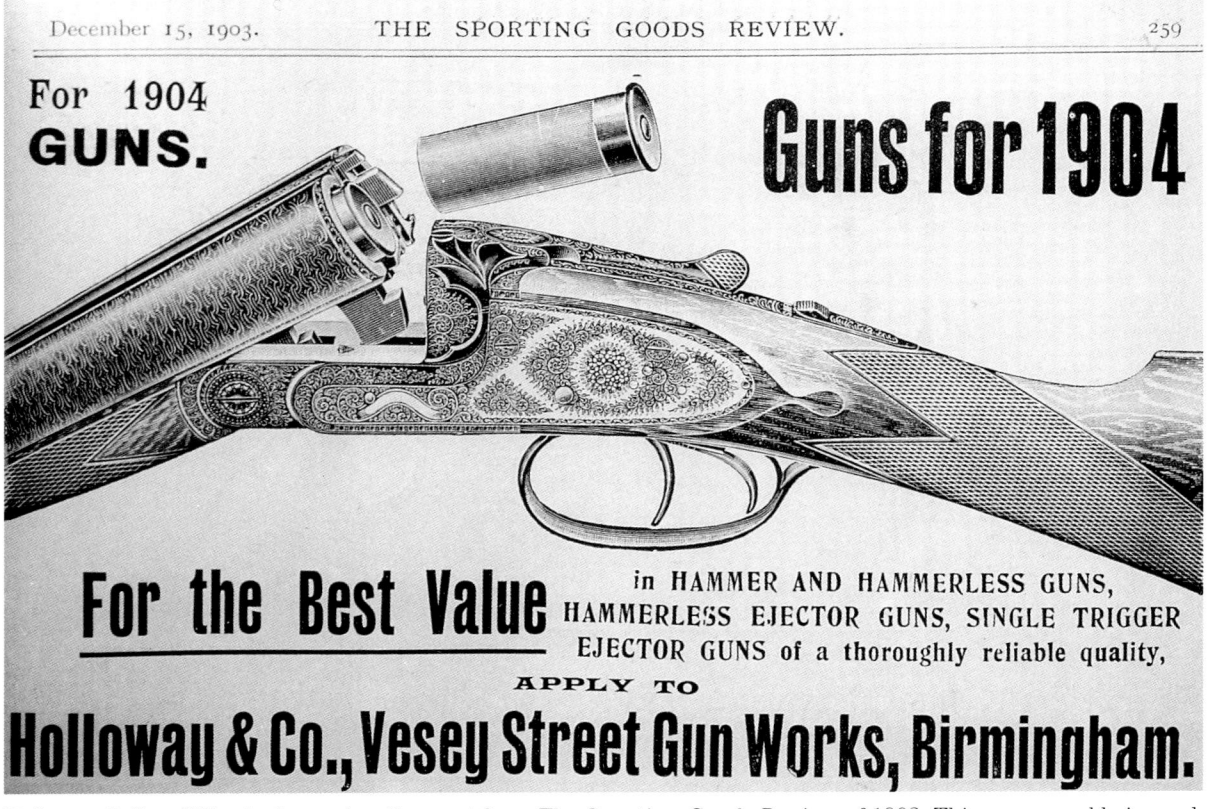

Guns for 1904

For the Best Value in HAMMER AND HAMMERLESS GUNS, HAMMERLESS EJECTOR GUNS, SINGLE TRIGGER EJECTOR GUNS of a thoroughly reliable quality,

APPLY TO

Holloway & Co., Vesey Street Gun Works, Birmingham.

Holloway & Co. of Birmingham advertisement from *The Sporting Goods Review* of 1903. This was a weekly journal for the trade, a rival to *Arms & Explosives*. Notice the blank space for the maker's name

fashionable London thoroughfares have always commanded high rents and, inevitably, these costs have ultimately to be borne by the purchaser of the goods from such shops. It follows, therefore, that one way of offering a gun, or indeed any article, at a more competitive price without in any way compromising the quality of the goods, is to trade from a less fashionable location.

For instance, in his 1908 catalogue, explaining the reason why his guns were competitively priced, Charles Boswell tells us that his rent for the first floor premises over 126 The Strand was but one sixth of those on the ground floor. Yet Boswell still had the advantage of a good address, not the most fashionable location, but still a good one.

There were, in fact, several gunmakers specialising in competitively-priced guns in what we might call peripheral central London locations. To pick a couple of names at random to illustrate this point, W.J. Jeffery was in Queen Victoria Street and The Army & Navy Stores could be found in Victoria Street.

The provincial retailer would also have enjoyed lower cost premises. True he lacked the London cachet but, especially if the business was long established, he would have benefited from local pride and loyalty with succeeding generations of sportsmen dealing with succeeding generations of gunmakers in the mellowing premises.

The important fact to realise is that no matter what his address, a gunmaker still had access to the retinue of out-workers who composed 'the trade'. So, for a special commission, he could go to exactly the same men who made the guns that were retailed by the most prestigious names in the business.

The problem facing a gentleman of moderate means, even working within Mr Walker's

A GUN FOR A GENTLEMAN

Charles Boswell of London bar-action sidelock ejector. A gun such as this, costing fifteen or twenty guineas less than a West End gun but probably made by the best London out-workers, represented splendid value for money

Thomas Horsley of York back-action sidelock ejector gun, line engraved only

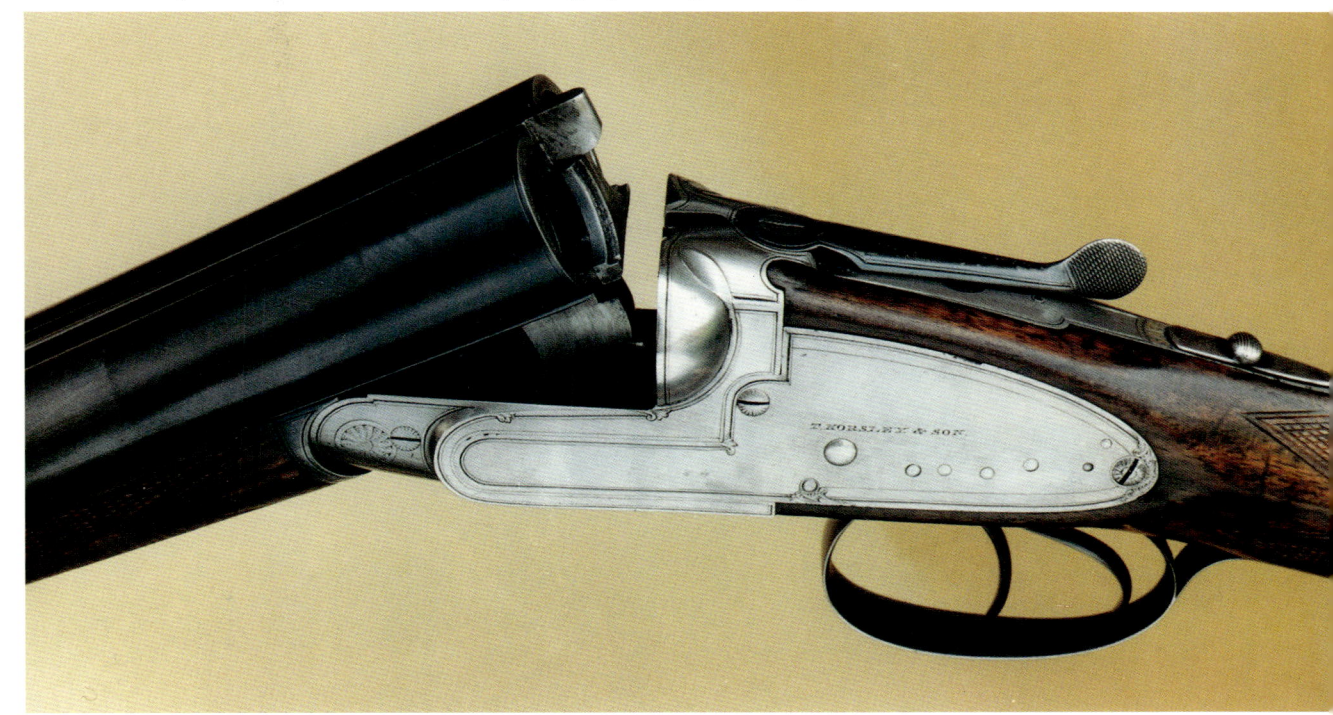

W. R. Pape of Newcastle. A page from his catalogue of the mid-1890s

A. Hill & Son of Horncastle bar-action sidelock 12-bore ejector

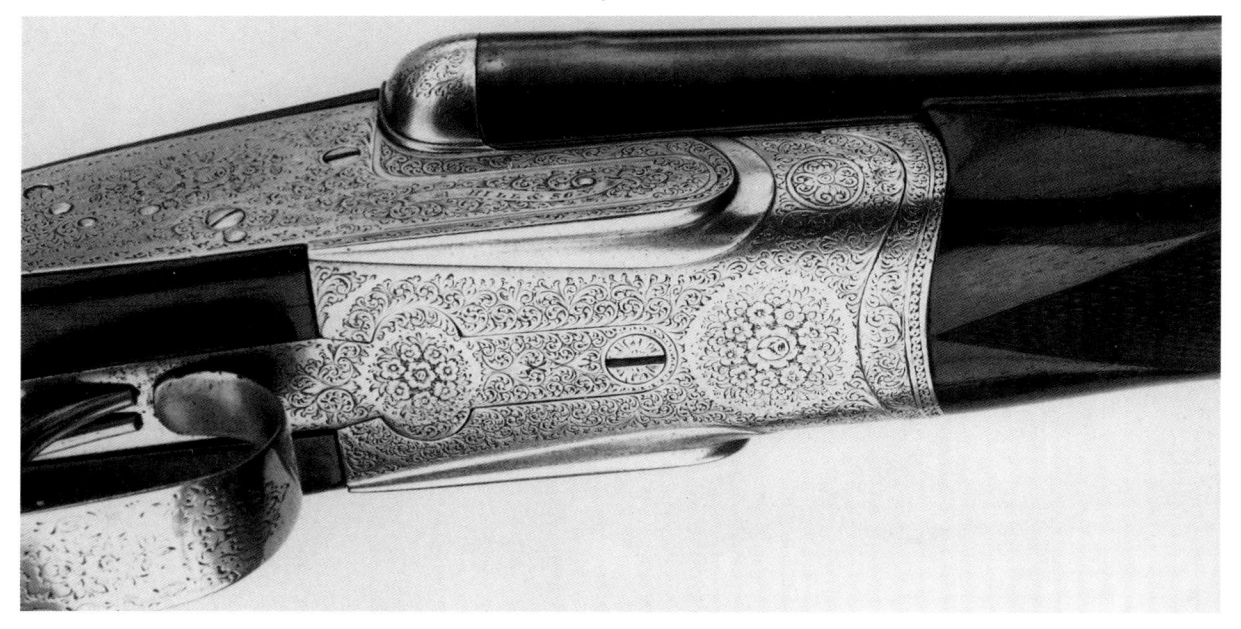

Page from Nobel advertising brochure *c.* 1905

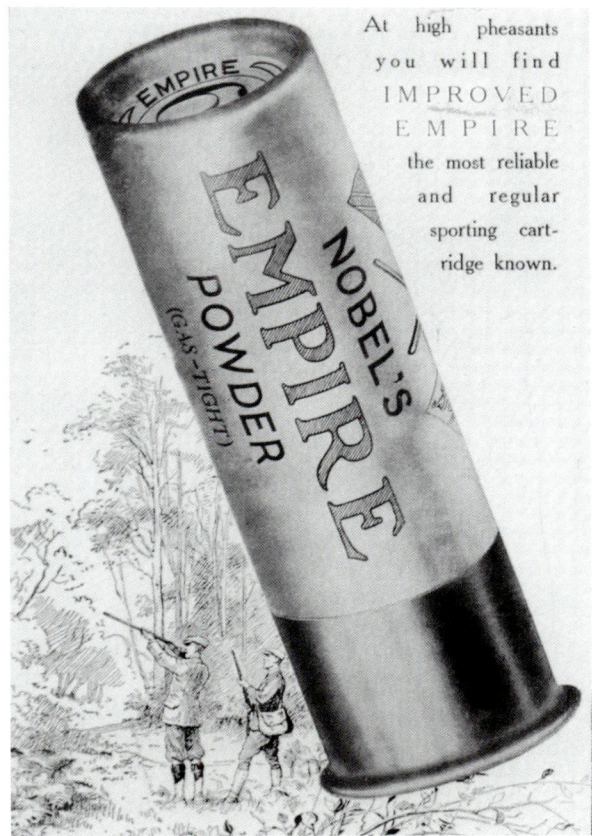

guidelines, was the apparent size of the choice confronting him. However, if he understood the way the gun trade was organised at that time, he would have realised that practically all these competitively-priced guns were made by relatively few makers in Birmingham, no matter which names were engraved on their ribs and locks or which proof marks they had stamped on their barrels and actions. The best clue that the discerning layman had to this state of affairs was in the products of those makers who used some sort of distinctive feature on their guns. The surprise is that a greater proportion of the sporting public looked no further than the vendor's name.

The most blatant example of this mode of trading had been with the W. & C. Scott hammerless gun that used no less than seven patented features, the rights to which were owned by the Scott firm, yet this gun was retailed by the greater proportion of the better end of the British trade. However, by the turn of the century, the Scott firm had merged with that of Thomas Webley to form Webley & Scott and the complex back-action Scott gun had been superseded. The gun that had replaced it was much more in the middle of the mainstream of fashion, being a Rogers patent bar-action sidelock.

The Rogers patent was just one of many

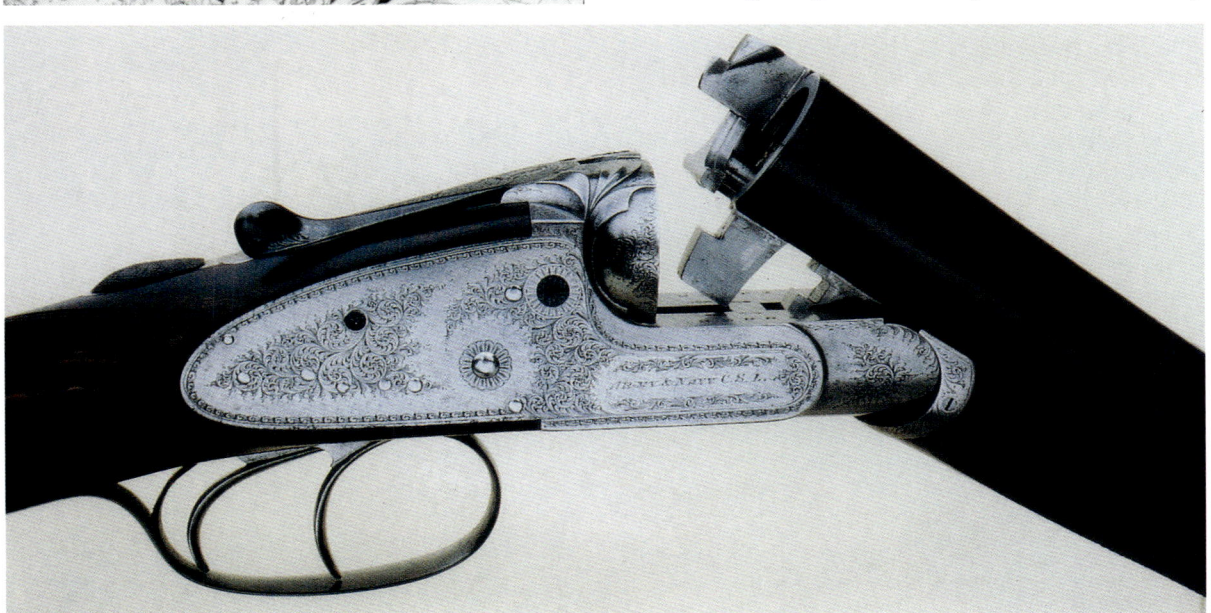

Army & Navy Company Stores Ltd gun "No. 4 Special Quality"

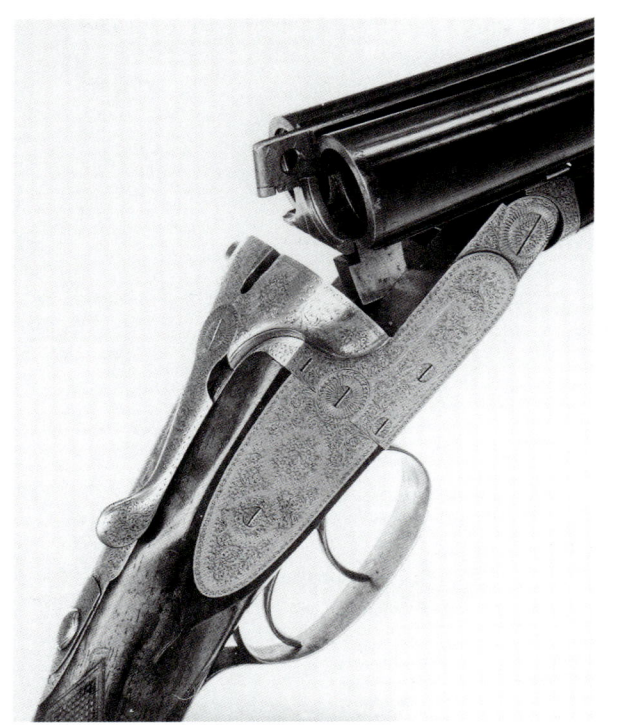

J. Lang of London boxlock ejector, fitted with side plates

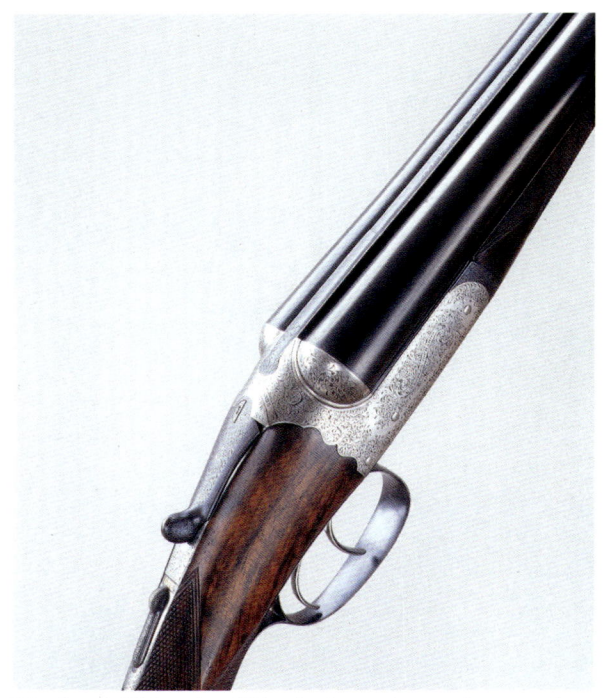

Westley & Richards of Birmingham. A superb example of the refined Anson & Deeley action

specifications which had been obtained in the 1880s for hammerless actions. It was simple, indeed as originally made it was seen as an action for modestly-priced guns towards the bottom of the market. The patentees, John Thomas Rogers and John Rogers, were part of a family who ran an action filing business in Lower Tower Street, Birmingham, but they sold the rights to their design in 1882 to the firm of J.P. Clabrough & Bros in Whittall Street, who had a thriving business exporting sporting guns to the United States of America and even their own retail outlet in California. The Clabrough factory also made actions for the rest of the trade. As a natural consequence of this diverse business, in terms of the range of guns both for the USA and those for other 'gunmakers', what had been a rather basic, even coarse, action, was refined, becoming both figuratively and literally polished.

Once the Rogers action became free of patent protection, other action makers could offer it to their clients and, since it is rare to find gun parts other than locks marked internally with the name

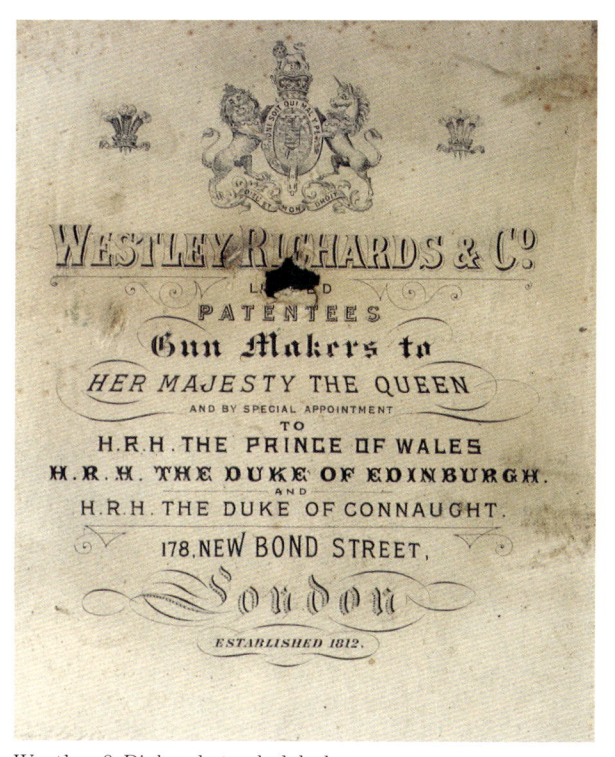

Westley & Richards trade label

A GUN FOR A GENTLEMAN

Left: Westley & Richards advertisement from *The Shooting Times*, 1898

of their makers, it is impossible to know where an action originated from.

An exception to this rule is the Webley & Scott, since that firm adhered to the use of the distinctive lever work that had been patented in 1882 by Thomas William Webley and Thomas Brain, an action filer, which was known as the Webley Screw Grip. This lever work was not only used on Rogers' sidelock guns but is also seen on a variety of grades of boxlocks and so all can be recognised as originating from the Premier Works, Scott's one-time factory in Lancaster Street, Birmingham.

The question of the selling price of a screw grip Webley/Rogers is complicated by the fact that a range of grades were on offer and, to a degree, the price was fixed by what the market would stand. However, a good guide is to be found in the Army & Navy catalogue of 1907, which lists no fewer than six grades with a finely graduated price range of nineteen pounds to forty-five guineas. The No. 6 model is described as 'good, sound and reliable', while the No. 1 was claimed to be 'very best quality and workmanship throughout, well

Below: W. W. Greener of Birmingham 'Facile Princeps'

balanced and shooting guaranteed'. Moreover, the claim was made that their guns were 'everyway equal to those supplied by West End firms at much higher prices'.

One of the most certain ways of ensuring that a value-for-money gun was sound was to choose a basic design that was simple to make. In 1900 this meant an Anson & Deeley boxlock. Even in an age when skilled craftsmen could only dream of wages of five pounds a week and consider themselves fortunate to earn three, the difference labour costs made to the final cost of a gun with an action that was simple build and stock was significant.

For those who wanted the best of both worlds – the looks of the sidelock and the price of the boxlock, there were boxlocks with false side plates. Some, like the Lang illustrated on page 30, were simply and honestly just that, but other guns were more sophisticated with dummy lock plates which had engraved upon them what purported to be all the pins visible on an ordinary sidelock. Naturally, one wonders just who was being deceived.

Pride of place in any consideration of the really good boxlock sporting gun must go to the firm of Westley Richards, then of Birmingham, Paris and London. After all, both Anson and Deeley were their employees and the firm maintained the Anson and Deeley patent for its full term of fourteen years until 1889. In this period, the Westley Richards firm defended the patent rights, and in so doing famously lost to W.W. Greener when they tried to claim too much. However, they reaped the rewards of their patent both in terms of the guns they made themselves and those actions that they licensed others to make.

By the turn of the century, the best Westley Richards Anson and Deeley guns had become more rounded and racy, in contrast to the rather boxy and angular look that characterised the guns made soon after the patent had been granted.

However, it is not for their looks that Westleys are admired by gunmakers. Right from the days of the percussion muzzle-loader, the guns of this famous house had been noted for their very restrained and limited engraving, stocks which had, in some cases, absolutely no swirling figure in the grain but with workmanship of the highest order, in fact, the epitome of understated quality.

The W.W. Greener gun which led to the legal battle between these two giants of the Birmingham trade was another body action gun, best known by the name its inventor gave it, 'Facile Princeps' or 'Easily First'. Incidentally, the Greener victory was regarded by the Birmingham trade as more a reward for legal legerdemain than a discovery of the truth. However, from our point of view, the interest in the Facile Princeps derives from the fact that there was a range of guns using this action within the price range suggested by Mr Walker. Like the rival Westleys, the Greeners were

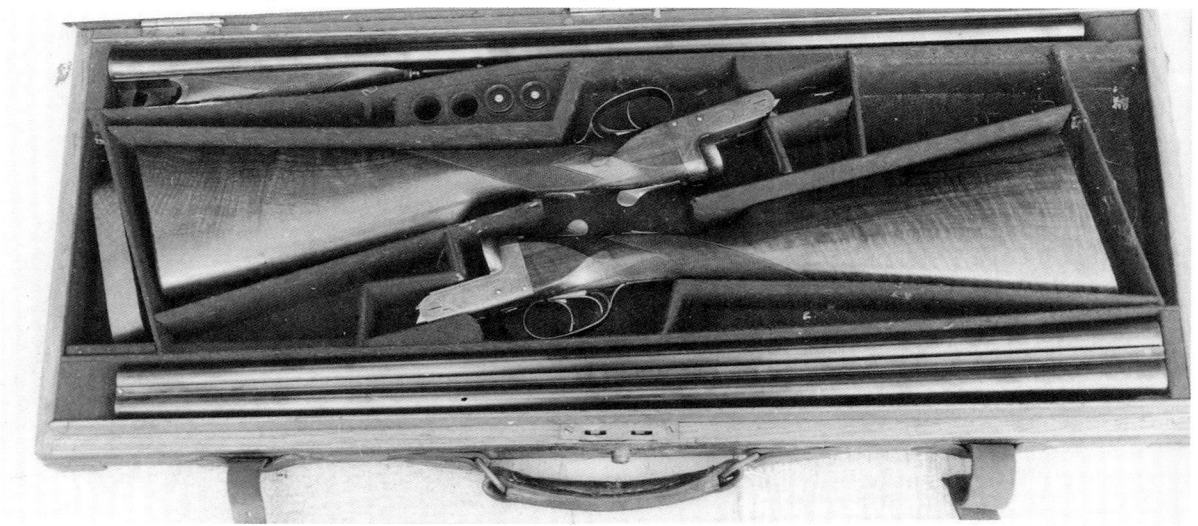

J. Robertson of London. A pair of 12-bore Anson & Deeley ejectors

CHARLES BOSWELL,

Gun and Rifle Maker,

126 Strand, London, W.C.

(First Floor).

LIST OF

Second-hand Guns, Rifles, Revolvers & Leather Goods.

Hammerless Ejector Guns

BEESLEY. Best quality 12-bore, side lock Hammerless Ejector gun, single trigger, carved fences, 30-inch Whitworth steel barrels, cost 75 guineas, in best case. Price **£35.**

PURDEY. Best quality 12-bore, side lock Hammerless Ejector, Whitworth barrels, condition equal to new ... ... Price **£55.**

DICKSON. Fine quality Anson and Deeley Ejector, steel barrels, finely finished, left choke, in excellent condition ... ... Price **£18.**

BOSWELL. Best quality 12-bore, side lock Hammerless Ejector, steel barrels, single trigger, left choke, gold oval, handsome stock, condition equal to new Price **£35.**

COGSWELL AND HARRISON. Fine quality Anson and Deeley Ejector Pigeon Gun, full choke both barrels, bored for 2¾-inch cases, gold oval, ½ pistol grip, fine condition, cost about £45 ... ... Price **£15.**

BONEHILL. 12-bore Anson and Deeley Ejector, 28-inch barrels, cross bolt, left choke, weight 6¼lbs., good condition ... ... Price **£10 10s.**

TOLLEY. Fine quality Anson and Deeley Ejector, 12-bore, 28-inch barrels, left choke, top lever, in good condition ... ... Price **£14 10s.**

Hammerless Ejector Guns (Continued).

BOSWELL. Good quality Anson and Deeley Ejector, steel barrels, top lever, left choke, an excellent killing gun, only used a few times, cost originally £20 Price **£13.**

PATRICK. 12-bore Anson and Deeley Ejector Pigeon Gun, steel barrels, full choke, good condition Price **£9.**

WOODWARD. 12-bore side lock Ejector, Whitworth steel barrels, a best quality gun, weight 6¼lbs., left choke ... ... Price **£45.**

LANG. 12-bore, best quality, side lock Ejector, Whitworth barrels (no engraving), left choke, fine condition, in leather case ... ... Price **£35.**

BOSWELL. Fine quality, side lock Ejector, Damascus barrels, left choke, top lever, in good condition, cost £35 ... ... Price **£20.**

BOSWELL. 16-bore Anson and Deeley Ejector Gun, cross bolt, 28-inch steel barrels, finely engraved, excellent condition ... ... Price **£16 10s.**

BOSWELL. 20-bore Anson and Deeley Ejector Gun, 30-inch steel barrels, ½ pistol grip, cross bolt, finely engraved, cost £30 ... ... Price **£18.**

BOSWELL. 12-bore Anson and Deeley Ejector, steel barrels, left choke, cross bolt, finely engraved and beautifully balanced ... ... Price **£12 10s.**

Charles Boswell's list of second-hand guns, from his catalogue of 1908

distinctive guns, but in a more flamboyant style. Whether this was the influence of some of the export markets or William Wellington's (W.W. Greener) preference, is an open question.

There is no doubt that these two firms remained rivals, competing for the same market with guns of the same basic specification, but which differed in every detail. The lockwork has already been noted and, while both used a trigger-bolting safety, Westleys used a conventional top slide, but Greeners used an unusual side safety on the head of the stock. It was not everybody's favourite, but it was another Greener patent.

Much the same is true of the ejector work, which were again in-house designs of the respective firms. Westley Richards has another Deeley patent which has been used across the trade. By contrast, the Greener, which was claimed as superior by its inventor on the basis of the few limbs involved and called by him 'The Unique', has other, much less flattering names bestowed upon it by those whose lot it is to attempt to repair it.

As for the two designs of top extension, both first appeared on pinfire guns when the breech-loader was new. Once more two fancy titles were given to these parts. The bulbous swelling on the Westley became the curiously-named 'Doll's Head', while Greener's 'Wedge Fast' for his cross bolt at least gave a clue to its function.

In competition with these two distinctive guns, practically the whole of the rest of the trade offered more or less standard Anson and Deeley boxlocks, of which the Lewis 'Gun of the Period' can be regarded as a good representative. They were by no means standard, having variations in the lever work and the ejectors, but minute consideration of these would be tedious.

More interesting, because they illustrate the importance of the market, are the conventional Anson and Deeleys retailed as less expensive alternatives to their best guns by Messrs Boss in St James's Street. It will be recalled that this firm always advertised as 'Makers of Best Guns Only'. To preserve this claim, the boxlocks were given a maker's name of 'J. Robertson', which was, in fact, entirely honest, since John Robertson was, at this date, the owner of the firm of Thomas Boss.

On top of all these options, the potential purchaser of moderate means would have been tempted by the offerings of those who dealt in second-hand guns. As with all functional mechanised items, considerable caution needs to be exercised when making a purchase and much depends on the provenance of the piece. After all, there is all the difference in the world between a genuine widow's sale and a wreck tarted up to sell. The problem facing the potential purchaser of a used shotgun in 1900 was that, because percussion caps used in contemporary cartridges left residues in the barrels, which encouraged rapid rusting, a neglected gun could have its barrels ruined in a couple of seasons and still look very much as new externally. Superficial rusting can, of course, be polished out of the bores of a shotgun and, if the barrel was originally well within its proof dimensions, it can remain in proof after the corrosion is removed. The danger was that a dealer might be tempted to lap out serious pitting and take the bore diameter beyond its permitted size. This doesn't necessarily render the gun dangerous but, to make sure of its safety, it should then be reproofed. This cost money and posed a degree of risk, so the temptation for the unscrupulous dealer was to pass it off onto an unsuspecting purchaser, who would be getting not just a poor bargain, but a potentially dangerous one to boot.

The proof laws exist to guard against this sort of malpractice, but laws are no use unless they are enforced and there were, it seems, very few prosecutions for selling out-of-proof guns. Another problem faced by the purchaser of a second-hand gun would have been the fit of the stock, which again could be remedied by expert attention.

The sum total of these pitfalls was that the purchaser of a second-hand gun either needed to be very lucky to find a gun in good condition that suited him as well as a gun made for him, or he needed honest expert advice. As they could guarantee neither of these, the authors of the time advised their readers to avoid the second-hand market. This, after all, was a simple, safe line to take.

Despite this sound advice, the second-hand gun could offer the gentleman who had his eye

SECOND-HAND EJECTORS.

Ejectors, Hammerless and Hammer Guns, by all the First London Makers at Half the Original Cost and as good as New.

303 Rifles, Martini, Winchester, and Lee Metford, Double ·450, ·500, and ·577 Bores in Stock at Low Prices.

REVOLVERS BY COLT, SMITH & WESSON, & WEBLEY.

Send for Catalogue, Gratis.

C. B. VAUGHAN,
39 STRAND, LONDON, W.C.
Telephone No. 5340 Gerrard.   Established over 100 Years.

Above: C. B. Vaughan of London, specialist dealers in second-hand guns, rifles and pistols, from *The Shooting Times* of 1898

on the price one huge bonus and that was the name engraved on the gun. If this was important to him, or he moved in circles where labels mattered, then he could have the precious cachet at an affordable price. On the other hand, if he were so minded, he could boast to those who had paid the maker's price about the snip of a bargain he had snapped up.

Notwithstanding Sir Ralph Payne-Gallwey's ignorant description of the Birmingham ready-made gun as a 'Monstrum Horrendum', which rightly earned him the ire of the Birmingham trade, they were splendid value for money. In reality, the best of the Birmingham guns could be carried and used with pride.

Below: The Charles Lancaster shooting school at Stonebridge, west London. An unknown shooter waits for a clay from the tower. (Photograph by courtesy of Mrs A. Thorn, from the family archive)

CHAPTER FOUR
A Gun for a Young Officer

In 1900, 'The Empire on which the sun never set' was defended by an army whose men expected to see the world and a navy that had to control the high seas. Inevitably, many of the young men who were officers in these forces were the scions of the sporting gentry and the opportunity for exotic sport was a prime attraction of service life. The guardians of the Empire represented a very important market for the gun trade. A gunmaker in a garrison town who saw the rotation of regiments with some being drafted overseas, or his counterpart in a port city where naval units came and went, had a flow of business to be envied by a market town trader with only a static farming clientele to depend upon.

Inevitably, there is a considerable overlap between the guns that form the subject of this chapter and those of the preceding one. For no matter where the gun was to be shot, the qualities looked for would be very similar.

Group of British officers with shotguns in India. (PHOTOGRAPH COURTESY OF THE NATIONAL ARMY MUSEUM)

A GUN FOR A YOUNG OFFICER

The emblem of The Army & Navy Company Stores Ltd

A page from Charles Lancaster's catalogue of 1893

Practically all the diverse outlets that made up the gun trade took a share of the service trade, from the specialist second-hand dealers such as Whistlers and Vaughans in the Strand to the more celebrated West End makers, some of whom offered plainer weapons, obviously with this market in mind. For instance, Holland & Holland had their No. 3 'Dominion' grade, while Lancasters offered both hammer and hammerless guns under the designation 'Colonial Quality' which was a registered trade mark. For a succinct contemporary quotation in the inimitable style of H.A.A. Thorn, it would be difficult to improve on the reproduced page from the Charles Lancaster catalogue of 1893.

It should be noted that this gun, in its basic form (that is a non-ejector with damascus barrels), retailed for less than half the price of the best gun offered at £60 in the same catalogue.

It is, of course, true to say that, for all markets, it is desirable that a gun should be reliable and inexpensive. However, for an impecunious young man about to set out for a remote part of the Empire, these two features assumed particular importance.

As can be seen from the Lancaster catalogue, to achieve the sort of prices that this market would stand, machine manufacture became essential. The problem for vendors was that to some potential purchasers, the term 'machine-made' conjured up the spectre of an ungainly, awkward lump of a gun. So we find phrases like 'semi-machine-made' in the advertising, apparently in the hope that purchasers would believe they were getting the best of both worlds. In the best examples this was indeed the case.

The firm of Cogswell & Harrison was prominent in the movement toward medium-priced guns and, as a result, claimed to be 'The Largest Manufacturer in London of Sporting Guns & Rifles'. They had a factory in Gillingham Street, close to Victoria Station, where they made both firearms and bicycles and offered a range of guns from ten to fifty guineas. Interestingly from our present perspective, they had one model, a side-plated boxlock ejector which they called the 'Sandhurst' gun. 'Exceptional value, many thousands sold. Unqualified satisfaction as testified by testimonials from all parts of the world,' proclaimed their advertisements.

The man probably most closely associated with the machine-made, lower-priced shotgun was C.G. Bonehill. Bonehill's philosophy was to make guns

Advertisement for The Victoria Small Arms Company, from *The Shooting Times*, 1899. This was a brief flirtation by Cogswell & Harrison with a lower part of the market

Heel plate of the Interchangeable gun

using the methods of the machine shop and to retail them directly to the customer. Part of this strategy consisted of inviting potential customers to tour the works in Belmont Row and, to facilitate this, the Bonehill catalogue included a small map of central Birmingham showing the works in relation to New Street railway station. Naturally, such a good idea was copied by rivals in the same market, notably the Midland Gun Company who, in 1902, moved to their Demon Works in Vesey Street from their previous address in Bath Street.

A perceptive part of Bonehill's selling methods was to choose the name 'Interchangeable' for his product, thereby inferring that any repairs could be carried out without the aid of a skilled gunsmith.

While this laudable objective may well have been Bonehill's original intention, in reality and over a period of time, the name 'Interchangeable' became more akin to a brand name and was applied to a complete range of shoulder arms. There were double shotguns, rifles, combination ball and shot guns, and what were called 'ball guns', which were double-barrel, smooth bores designed to shoot solid ball and all of these were available in both external hammer and hammerless forms.

To complicate the question of interchangeability even further, there were four grades of both hammer and hammerless shotgun, plus a heavyweight gun for 3-inch cases. Eventually, the name came to be applied to straightforward Anson and Deeley boxlocks.

So, while there were many parts that might appear to be common to the various styles of gun, contemporary tests proved that they could not be interchanged without fitting. To be fair, some of

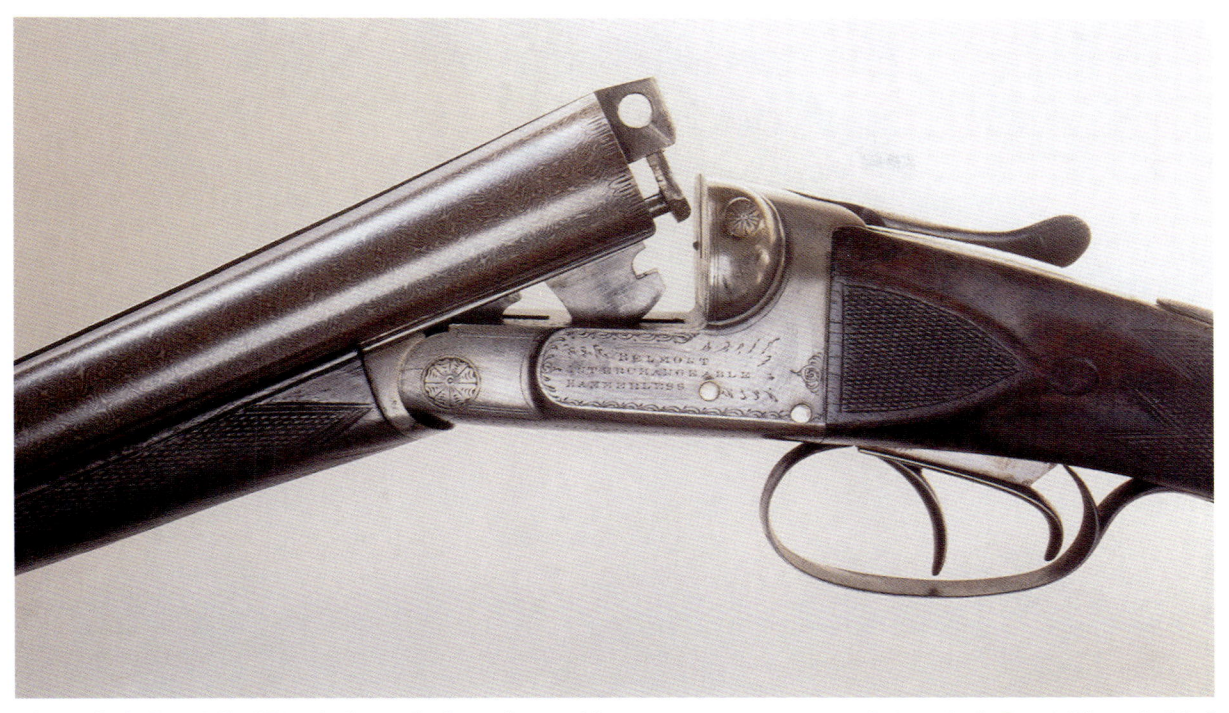

Above: C. G. Bonehill of Birmingham: the Interchangeable gun

Below: C. G. Bonehill's trade label

28 BORE HAMMERLESS
LADIES' AND YOUTHS' GUNS.

This make of Gun can also be made to Order in 16 and 20 BORES.

12 BORE GUNS OF THIS QUALITY ARE USUALLY KEPT IN STOCK.

These Guns are manufactured in Liege, Belgium, to our design, and according to our instructions. When London proved, we engrave OUR NAME AND ADDRESS on the Barrels, and the words "MADE IN BELGIUM," to comply with the Merchandise Marks Act.

PRICE ... £6 10 0

DESCRIPTION.

28 Bore Top Lever Treble Bolted Anson & Deeley Action, with Top Crossbolt Extension, 28in. Steel Barrels, Case-hardened Action, Anson Pattern Forepart Fastener, London proved, named **W J Jeffery & Co.**, Automatic Safety Bolt, Left Barrel Choked, Walnut Stock, no engraving. Weight, about 4½ lbs.

28 Bore Guns are usually kept in stock ready for immediate delivery, and also 12 Bores of the same pattern, at the same price.

16 and 20 Bore Guns are made to order, and can be usually manufactured in three or four months after receipt of order.

During the last few years, owing to cheap labour and an extensive use of machinery, it is possible to manufacture sound and reliable cheap guns in Belgium at some thirty to forty per cent. lower than they can be produced in England. In guns costing £12 and upwards, the English workman can still produce a superior weapon, but, in the cheaper grades, he is totally beaten.

The Belgian workman puts in twelve hours a day, as against the English workman's seven or eight, with the result that he produces fully forty per cent. more. His wages are generally twenty to thiry per cent. less per hour. Probably the highest testimonial that the Liege Manufacturers ever received is that quite recently the Birmingham Gunmakers sent a deputation to study their methods of production, and to report generally on gun manufacture in Liege.

ALL ORDERS MUST BE ACCOMPANIED BY CASH

the fine tuning needed was probably within the skills of a service armourer, so we should not completely dismiss the concept of the 'Interchangeable'.

Leaving aside the question as to whether guns were truly interchangeable or not, there is no doubt at all of the excellent value these guns represented. The price structure began at £8/10/0d. (£8·50) for a plain hammerless gun, the grade A, and rose to £15 for the grade D. This latter had 'ornamentation of the most artistic nature', consisting not only of extensive engraving but fancy border chequering as well. For external hammer guns, the prices ranged from £6/0/5d. (£6·02) for the plain gun up to £13 for a grade D. Incredible as it may seem, these prices were not the bottom of the market. Below them we find a whole new set of options in the imported gun. That Bonehills with their factory could be undercut dramatically by these imports, demonstrates the problems besetting them in the export markets of the world.

In the Jeffery catalogue of 1904–5, there is on offer a range of Belgian hammerless guns which are £2 cheaper than the cheapest comparable Interchangeable. The rival British trade would have tried to dismiss these guns as 'Belgian rubbish' and equate them with the dire hammer guns we will encounter later. In truth, this is not a fair assessment, certainly these Jeffery imports were adequately sound. Moreover, the Belgians were astute producers of guns to the style their export customers sought. Over the years all sorts of firearms had been made in Liege from Balkan-style pistols, via Moorish long guns, to Kentucky rifles in response to the needs of their markets. So it was perfectly natural for the Liegeois maker to produce these boxlocks '*à l'Anglais*' and thereby come up with a product very tempting to the young officer with a shallow purse and uncertain of the sporting prospects of his posting.

The over-riding difference between shooting in

Opposite page: A page from W.J. Jeffery & Co Catalogue of Season 1904–1905

Holland & Holland Paradox gun

the British Isles and the colonies was the variety of game that might be encountered. Moreover, some exotic beasts were potentially very dangerous and clearly a hunter needed more than a charge of birdshot at his command. From the earliest times, spherical lead bullets had been fired from smooth bore guns. The advent of the choke bored barrels with their constricted muzzles complicated, but did not preclude, the use of solid ball, despite the proof mark current between 1875 and 1887 which proclaimed 'Not for Ball'.

From the hunter's point of view, the problem of firing ball ammunition from a shotgun was simply that the shooting was erratic and hence the practical range was limited. This situation prevailed until 1886 when, after a protracted series of classic experiments, Lt-Col George Fosbery, V.C., of the Bengal Army evolved a system in which the choke portion of a shotgun barrel was rifled. By this means, with suitable bullets, it was found possible to extend greatly the accurate range of a shotgun firing ball while, at the same time, retaining a barrel that made excellent patterns with small shot. Interestingly, the influence of the rifling can be seen as a halo-like band of shot that forms around the pattern.

In fairness to other inventors and gunmakers, it should be noted that there was nothing new in the idea of partially rifling a barrel. There are guns by Joe Manton that exhibit this feature dating from the beginning of the nineteenth century and, as is the way of such things, the idea was probably very much older than Manton. It is also worthy of note that Fosbery seems to have envisaged riot control as an important role for his invention.

Be this as it may, Fosbery's experiments lead to his obtaining a British patent in 1885 which was worked by the London gunmakers, Messrs Holland & Holland of New Bond Street and called by them 'The Paradox'. That this gun was publicly endorsed by Sir Samuel Baker, one of the greatest and most influential explorers and adventurers of the period, is but one indication of its practical value. Hollands built guns of all gauges between eight and twenty-eight on the system and, once the patent had expired, the idea was widely used in the trade under a variety of fanciful names, amongst them: Moore & Grey's 'Anomaly', Bland's 'Euoplia', Tolley's 'Ubique', Cogswell & Harrison's 'Cosmos',

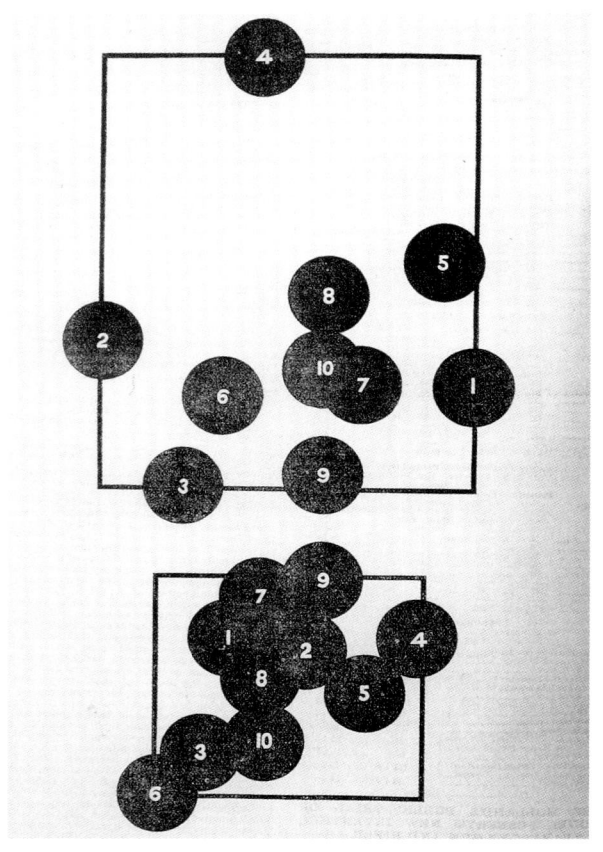

Pattern shot with Paradox gun at fifty yards, lower diagram, and one hundred yards, upper diagram

Lang's 'Afrindia', Army & Navy's 'Jungle' and Jeffery's 'Shikari'.

A widely advertised contender for the same market was the Lancaster 'Colindian', which sought to achieve the same results by different means. This was the application of the Lancaster oval bore, the invention of Mr Thorn's predecessor, Charles William Lancaster. It had been used on weapons as diverse as high velocity rifles and huge siege guns.

The prices of these ball and shot guns covered practically the whole range. At the top end a Holland & Holland Royal grade 'Paradox' would set the purchaser back sixty-five guineas and, therefore, was more likely to be bought by an Indian prince than a young subaltern. At the bottom of the market, in 1901, the firm of C.W. Andrews Ltd, of Great Winchester Street, London, with works at 13 Bath Street, Birmingham,

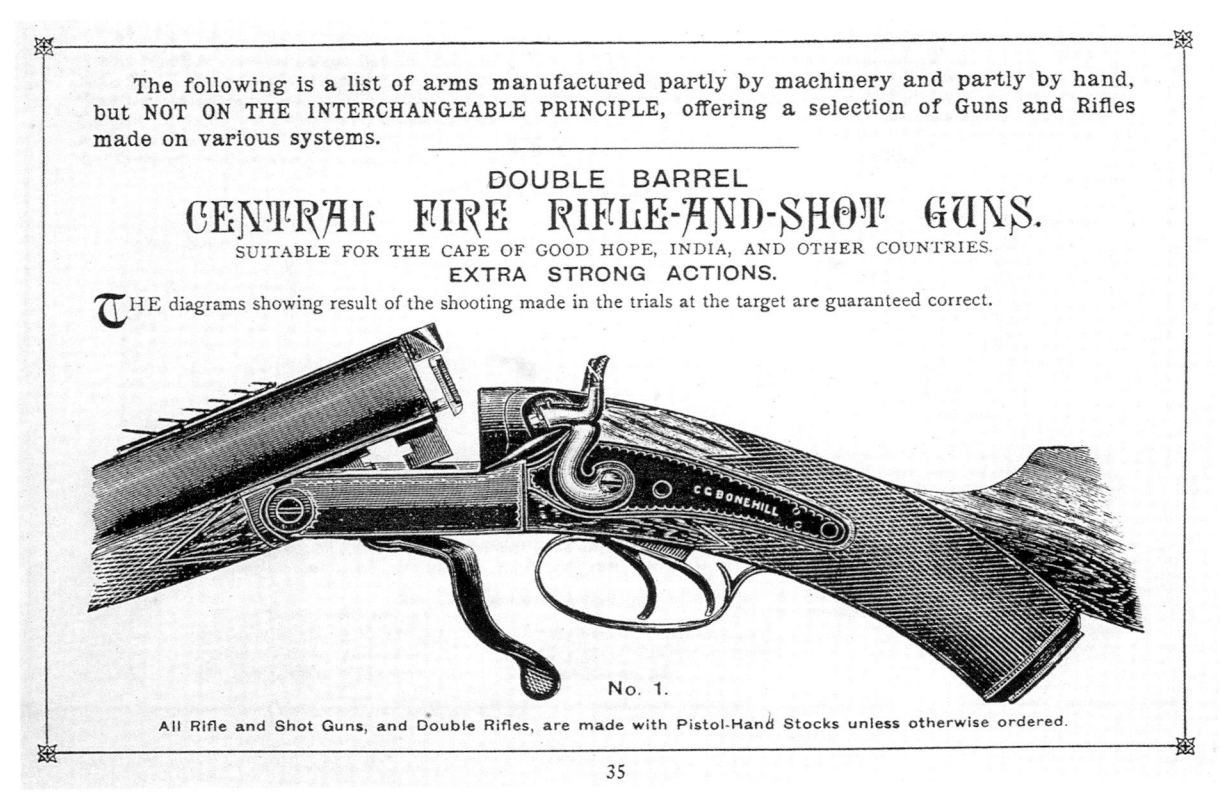

> The following is a list of arms manufactured partly by machinery and partly by hand, but NOT ON THE INTERCHANGEABLE PRINCIPLE, offering a selection of Guns and Rifles made on various systems.
>
> ## DOUBLE BARREL
> # CENTRAL FIRE RIFLE-AND-SHOT GUNS.
> SUITABLE FOR THE CAPE OF GOOD HOPE, INDIA, AND OTHER COUNTRIES.
> ### EXTRA STRONG ACTIONS.
> THE diagrams showing result of the shooting made in the trials at the target are guaranteed correct.
>
> No. 1.
>
> All Rifle and Shot Guns, and Double Rifles, are made with Pistol-Hand Stocks unless otherwise ordered.

A page from C. G. Bonehill's catalogue of *c.* 1892, showing the firm's ball and shot guns

advertised a ball and shot gun for £7/10/0d. (£7·50). This gun did not have the benefit of an exotic name and a fanciful title evidently increased the value if the Army & Navy 'Jungle' is anything to go by. Prices started at £11 but not until the £25 mark was 'accuracy guaranteed'. On the subject of cost, Jeffery's price for 100 12-bore cartridges loaded with black powder and ball was 14/6d. (72½p), in other words almost double the price of shot cartridges.

One option open to the impecunious purchaser to increase the versatility of an armoury was to buy one or more Morris tube sub-calibre inserts to enable either his shotgun or larger calibre rifles to fire one of a variety of lower-powered small rifle or pistol cartridges.

The Morris tube had originated back in the 1880s, as a device to enable the military to practise with their rifles without the need to waste time marching out to distant ranges where their large calibre, powerful rifles could be fired safely. Instead, a small calibre rifle barrel was fitted inside the service rifle barrel. Through this was fired one of the smallest centre fire rifle cartridges ever made at a cleverly designed target. This, at a trifling cost for the ammunition, enabled rifle practice to take place in the drill hall using the sights intended for the service ammunition.

This concept had obvious civilian potential and the Morris Tube Company promoted their wares vigorously, witness the range of calibres that could be fired and equally the variety of sizes of gun and rifle that could be used. It has to be admitted that sighting problems would limit the usefulness of a combination of a small bore rifle cartridge fired through a conventional shotgun devoid of a backsight. However, at shorter ranges, the combination would still be effective, while, in a rifle, the full potential of the ammunition could be realised. So, at thirty-five shillings (£1·75) the tube could be the equivalent of a rook rifle without the

MORRIS TUBES FOR SHOT-GUNS,
AND EVERY DESCRIPTION OF
GUNS & RIFLES, Sporting and Military.

"It is not generally known that the Morris Tube can now be adjusted to shot-guns without any alteration or chance of damage to the gun. Thus, owners of 10, 12, 16, or even 20 bore guns can have tubes fitted to their weapons, instantly converting them into combined rifles and shot guns, which, for rabbit and rook shooting will be found very useful—the rifle barrel or tube for stalking, and the shot barrel for flyers and runners.

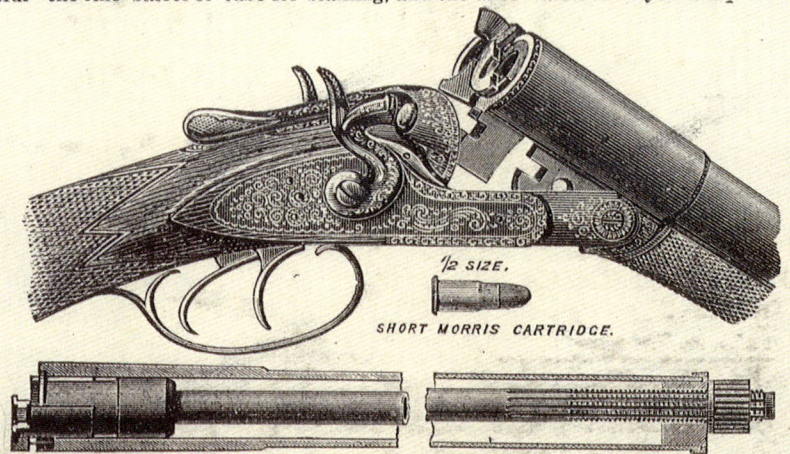

½ SIZE.

SHORT MORRIS CARTRIDGE.

When once the tube has been fitted it can be placed in, or taken out of, the gun in a few seconds, and is so constructed that it cannot be fixed in any position but the correct one. The illustration will show the position of the tube in the barrel. Tubes of any calibre, from ·297 to ·44, can be fitted at a cost of something like 35s. The tubes are adjusted to shoot point blank at 80yds., but beyond that distance a rear sight is necessary. Of course a little additional weight is added to the gun, and the balance somewhat disturbed, but the convenience of having a gun or rifle at will at such a small outlay, compensates for many minor disadvantages. We have shot with a weapon fitted with a tube, and found it handy and accurate."—*The Sporting Times*, June 29, 1895.

Smooth bore tubes also are made, and fitted to an ordinary gun, and are most valuable for shooting small birds, as they at once reduce the bore to that of a collector's gun. These tubes also are capital things for lads learning the use of the gun.

Tubes can be fitted to **Guns and Rifles** of almost any calibre, including the new ·303 bore Magazine Rifle, and used with the Morris ·297/·230. Long and short non-fouling ammunition accuracy is guaranteed. Care should, however, be taken to use only ammunition that is packed under the Company's registered Bull's-eye label with Bull's-head centre.

ROOK AND RABBIT RIFLES, DOUBLE GUNS,
and every description of Shooting Requisites.
MINIATURE MARTINI RIFLES,

specially made for Morris ammunition for use by Cadets at Public Schools, &c.; also in various patterns for sporting purposes. These rifles are extremely accurate and powerful, and are of best workmanship and high finish.

Fully Illustrated and Priced Catalogues can be obtained free on application to the

Morris Tube Ammunition and Safety Range Co., Ltd.,
11, HAYMARKET, LONDON, S.W.

THE TRADE SUPPLIED.

C. G. Bonehill 12-bore hammerless ejector. This gun uses the same action as the Interchangeable, but is built to a much higher standard and finished to weigh just under six pounds

Brass converter for 12-bore

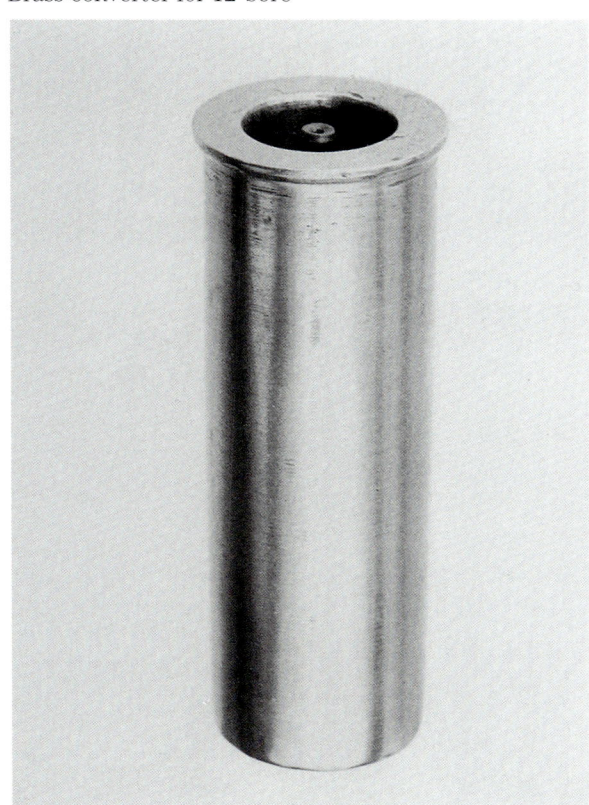

bulk of a complete extra firearm.

In a similar vein to the Morris tube, another item that could usefully be tucked into the young officer's gun case was a few everlasting cartridges. These brass or steel chamber inserts were not a new idea, indeed they go right back to the beginning of the breech-loader. Their purpose was to enable basic muzzle-loading components – caps, powder, wads and shot, to be assembled in such a way that they could be used in a modern breech-loader.

The value of such a device was not for large-scale shooting, but to fill the pot when supplies of ready-loaded cartridges were unobtainable. For, by contrast with the more limited distribution of loaded ammunition, gunpowder, albeit of variable quality, was much more readily available. Likewise shot, which, in any case, it was possible to improvise.

With this last touch, our young officer could congratulate himself on being fully equipped and look forward to good sport in foreign lands.

Opposite page: The Morris Tube Ammunition and Safety Range Co. Ltd. Advertisement from H. Sharp, *Practical Wildfowling* (1895)

CHAPTER FIVE

A Gun for a Gamekeeper

A shotgun on the arm of his tweed jacket was close to being a badge of office for a gamekeeper. The gun that was the keeper's almost constant companion was, more than any other, a working gun. As is the way of such things, there was no one standard gamekeeper's gun, but there was a range that spanned practically the complete spectrum from the very best to the dire.

How a keeper came to have a best gun represents a snippet of social history that throws a revealing light on the times. On many estates when the squire bought a new gun he passed his old one on to his head keeper. He, in turn, passed his current gun, an earlier hand-me-down from his master, to one of the under-keepers.

This process was only a part of an ethos which held that things had to be worn out before they were discarded. What had been 'Sunday best' suits ended up as working clothes. Where cottagers had sheets that had worn thin in the middle, they split

W. Burgess, head keeper at Escrick Park, Yorkshire. Sidney Smith.
(PHOTOGRAPH BY COURTESY OF THE YORKSHIRE MUSEUM)

Chamberlin & Smith, Norwich. Game food advertisement from *The Shooting Times*, 1898

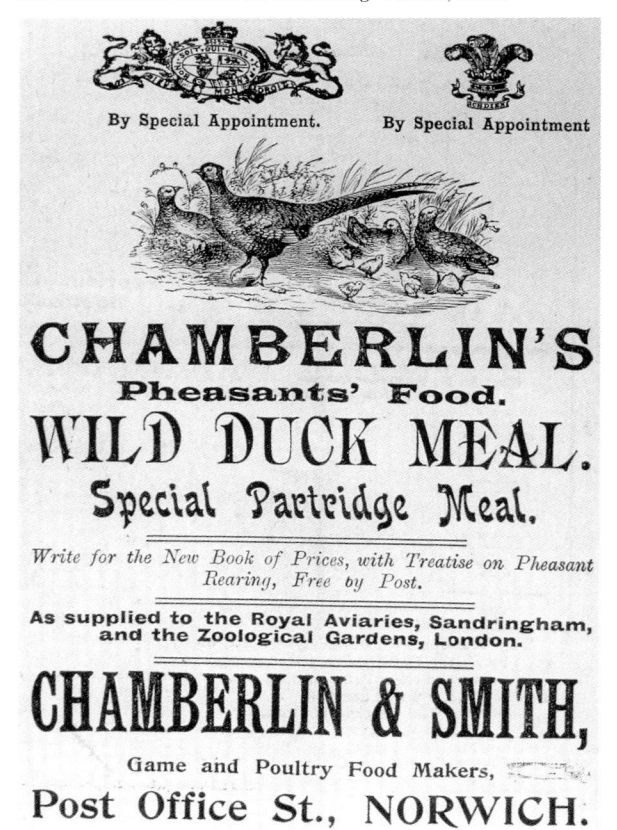

A GUN FOR A GAMEKEEPER

James Purdey of London best hammer gun, built 1871. Chequering completely worn away by gamekeeper's use

W. R. Pape of Newcastle. Page from catalogue of mid-1890s

THE "KEEPER'S."

£6 10s.

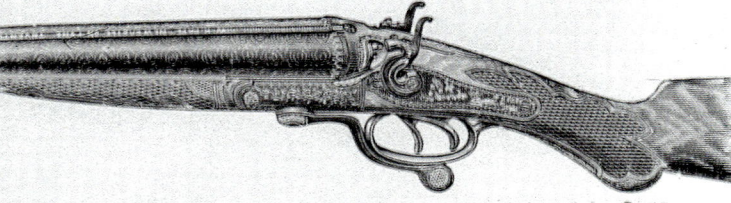

Undoubtedly the best value ever produced in plain Guns.

12, 16 or 20 Bore—Fine Damascus Barrels—Left Choke—(both Choke if desired, no extra charge)—Dolls head extension Rib—Edge's Patent Forend Lifter—Back-action and Rebounding Locks—With or without Pistol Handle Stock, made and kept in Stock, with both Top and Double Grip Levers (see cut)—Clean finished handsome Walnut Stocks. latest patents and improvements—Every Gun shot at our own ranges and guaranteed.

Price £6 10s.

HUNDREDS OF TESTIMONIALS received about this Gun, same as given here.

"Redesdale Cottage, Otterburn, Northumberland, February 7th, 1888.
"Dear Sir,—The new 'Keeper's' Gun arrived quite safely, its shooting powers are first class. My Gamekeeper (Mr. Stobie) likes it exceedingly well, and says its a regular "Nailer"—which is very satisfactory. I enclose you cheque value £6 : 10 : 0.   "I am, Sir, yours respectfully, WM. HODGSON."

"Killarney House, Killarney, Ireland, February 25th, 1889.
"Dear Sir,—Please find enclosed cheque £6 : 10 : 0, the price of the 'Keeper's' Gun. It arrived safely on Saturday morning. I consider it a *very handy*, hard shooting, and well-finished gun for the money—and would, without doubt, sell in Ireland, at £3 more. My friends are all well pleased with it. I am just off to the mountains to get some woodcocks.
"Yours, &c., T. DAVIDSON."

HOW TO CARRY A GUN.
From "Shooting," Badminton Library, Messrs. Longmans, Green & Co.

WINNER OF THE GREAT LONDON GUN TRIALS HELD IN 1858, 1859, 1866 and 1875.

THOMAS BLAND & SONS,
GUN MANUFACTURERS, MERCHANTS,
AND MAKERS OF EVERY DESCRIPTION OF
Military & Sporting Breech & Muzzle-Loading Guns,
RIFLES AND REVOLVERS.

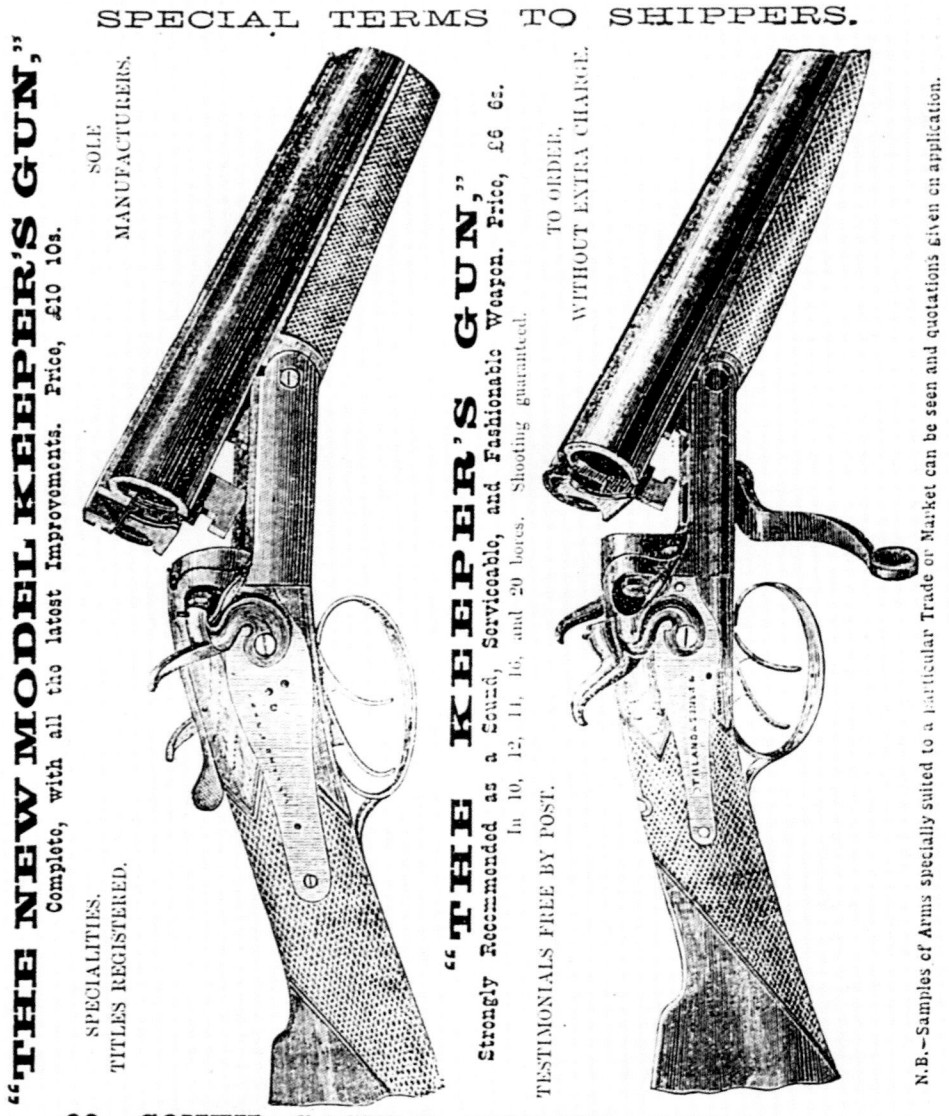

SPECIAL TERMS TO SHIPPERS.

"THE NEW MODEL KEEPER'S GUN," Complete, with all the latest Improvements. Price, £10 10s. SOLE MANUFACTURERS. SPECIALITIES. TITLES REGISTERED.

"THE KEEPER'S GUN," Strongly Recommended as a Sound, Serviceable, and Fashionable Weapon. Price, £6 6s. In 10, 12, 14, 16, and 20 bores. Shooting guaranteed. TO ORDER, WITHOUT EXTRA CHARGE. TESTIMONIALS FREE BY POST.

N.B.—Samples of Arms specially suited to a particular Trade or Market can be seen and quotations given on application.

62, SOUTH CASTLE STREET, LIVERPOOL;
106, STRAND, LONDON; & WORKS, BIRMINGHAM.

Opposite: Thomas Bland advertisement, from *Land & Water*, 1897

them in half and sewed the outside edges together to give them a new lease of life.

There is no certain way of recognising a best gun that has followed this route, but a strong clue is a stock so worn that the chequering is almost totally removed. Such wear is the result of years of handling by hands that are frequently soiled with earth from digging out rabbits and setting traps for ground vermin.

That fine guns survived years of rough work is but another testimony to their original quality and proves that the products of the fashionable 'West End' were not mere baubles. If the keeper or his employer were, for some reason, buying a new working gun, they would almost inevitably seek longevity in the best gun at a lower price. To a degree, of course, this was an impossibility because the careful, unhurried craftsmanship with fine materials which created the soundness of the best gun were reflected in the high original cost.

Nevertheless, there were guns marketed as 'Keepers' Guns'. In some instances, this was no more than a meaningless, catch-penny brand name, advertising a cheap and more or less nasty gun that had a dog crudely engraved on the lock plate. Marketing ploys of this sort can be lightly dismissed.

Not so the range of guns they set out to ape, the respected Thomas Bland 'Keeper's Gun'. While most lower-priced guns could be described as imitations of the very best guns, because this was what the gun-buying public demanded, the Bland philosophy was different, a no-frills approach, in which unseen quality was given greatest priority.

Therefore, there was no engraving, not that engraving was particularly expensive – total coverage with ordinary quality scroll engraving only added about £5 to the price of a high-grade gun. Rather the lack of engraving immediately conveyed the fact that here was a different set of priorities.

The two most common failings of inferior sporting guns of conventional design is that they

Thomas Bland of London, Birmingham and Liverpool, 'The Keeper's Gun' advertisement from *Gore's Liverpool directory*

Thomas Bland: The Hammerless Keeper's gun

Inset: Thomas Bland: The Hammerless Keeper's gun. Detail of heel plate

A GUN FOR A GAMEKEEPER

Thomas Bland: The Hammerless Keeper's gun. Detail of rib

'shoot loose' and that the lockwork wears to produce variable trigger pulls. Wear is inevitable in all instances where moving parts are involved. Long life is merely the product of reducing that wear or replacing the worn parts. The effects of wear can be reduced by careful design and workmanship on the one hand, and on the other by the use of metal of the right degree of hardness.

None of this was discovered by Blands or, for that matter, first used by them. The Bland innovation was to couple careful jointing and properly hardened steel lock parts with advertising which was lavish by the standards of the day. This, it was hoped, would remove the problem of there being no way a prospective purchaser could tell if the tumblers were properly hardened or judge the fit of a hinge pin.

Of course, the practical advantages of the Bland 'Keeper's Gun' were not limited solely to gamekeepers. Letters to the sporting press testify to the fact that others who might have been expected to buy a more expensive gun were among its purchasers.

In the same vein is the fulsome write-up that the Bland received in *How to Buy a Gun* by Bazil Tozer and H.A. Bryden. The latter contributed the section on the rifle and rifle shooting with, it must be noted, considerable fervour, generated by the

George Smith of Norwich. Game food advertisement from *The Shooting Times*, 1898

painful lessons learned by the British Army in the Boer War. On the subject of keepers' guns, Mr Tozer first of all condemns as 'scrap iron' some of the guns sold under this title, but carries on with an anecdote concerning the Bland gun that bears repetition.

> It is a remark that was made to me not long ago by a very wealthy man, who has never in his life, I believe, handled any but 'Keeper's' guns – and I must in justice add that he handles them to some purpose. 'I get a new one every few years,' he said, 'and the advantage I find is that you never need clean them or bother about them in any way. I am dreadfully careless about my guns, as you know, and my men are worse. The gun I have here has, I fancy, not been cleaned since it came from the shop, two or three years ago, and I know it has never been overhauled. I daresay the locks are in as shocking a condition as the barrels – the right hammer squeaks a little when I cock it; but next season I shall give this gun to the keeper and get myself a new one. When he has done with it he gives it to the gardener, and when the gardener has done with it he sells it. I don't believe in your fads about balance, and fit, and special mechanisms, and all the rest of it; and as for a hammerless, I wouldn't shoot my worst enemy with one. I pay six guineas apiece for these guns, and I get a new one about every fourth season, so that at the end of fifty years I shall have paid only about as much as you would like me to pay for a single "first-class" gun, as you call it; and, I tell you frankly, I believe I shall have had better value for my money than if I had bought one of your "aristocratic" guns.'

His shoulder gun was probably not the only gun a gamekeeper would use. Some advocated that he had a pistol in his pocket to add to his authority in violent and sometimes fatal affrays with desperate poachers. Wiser council was that, in untrained hands, a pistol was almost as dangerous to its user as his adversary and, in any event, we are concerned here with shotguns.

A more usual gun in the keeper's armoury would have been an alarm gun. This could be any device fired by a trip wire and one way of contriving this was to butcher and cobble a conventional shoulder gun, maybe a relic beyond any other use, deemed unsafe to fire with a valued hand grasping the barrel, but good enough to nail to a post or lash to a tree. There was a certain poetic justice in using a gun abandoned by or recovered from a poacher in this way.

Much more satisfactory from a practical point of view was a purpose-made alarm gun designed to be left out in all weathers. It is a measure of the value of such a gun that there was a proliferation of these on the market.

Advertised for a period of at least two decades prior to the heyday, but somewhat expensive at ten shillings (50p), was the Horsley 'Registered Design' alarm gun. In common with several other alarm guns, it was designed for a pinfire cartridge, the built-in firing pin of such a cartridge being an obvious advantage. The most ingenious feature of the Horsley was to cast the weight that slid down the vertical stem to fire the cartridge as a flattish cone, something like a coolie's hat. This would deflect some of the rain but, even with this protection, to function reliably the cartridge

Thomas Horsley of York's Registered Design alarm gun

Unnamed, low grade hammer gun with back-action locks and Henry Jones screw grip action. This gun was the lifelong companion of a Yorkshire keeper, a sniper in The Great War; it was rare to see him miss

would need to be changed regularly and dried out on the hob of the kitchen range.

Inevitably not all gamekeepers carried a sound gun, some were armed little better than the poachers they pursued. Given the constant use the keeper gave his gun, the weaknesses of an inferior firearm were ruthlessly exposed and one response to this was home gunsmithing. The ways in which such repairs were made are calculated to scandalise a gunsmith but, in their crude way, they were triumphs of ingenuity with very limited resources and technical skills.

The gun that illustrates this section was the lifelong companion of a gamekeeper and has extensive problems. It was never taken to a professional repairer and we can see a wondrous selection of bodged repairs necessary because, when the gun was new it was probably just about the bottom of the barrel of English manufacture. Indeed, at first glance, I suspected that it was of Belgian make.

What gave this impression was the style of twist pattern on the barrels. This was known in its native Belgium as 'Tordus' and was the cheapest of an extensive range of twist and damascus barrels made by the barrel makers of the Vesdre Valley, south-east of Liege. The River Vesdre, which is a tributary of the Ourthe, runs westward through the northern edge of the Ardennes and it was this source of water power that led to the establishment of a series of barrel-making workshops. In 1896 they employed about 2000 men in the various stages of making twisted barrels and produced about 600,000 a year.

Since this gun has the distinctive 'Not for Ball' stamp of the British proof marks current between 1876 and 1887, we have a reasonably narrow indication of its date of manufacture. At this time, a pair of Tordus barrels would have been sold by their makers for between five-and-a-half and seven-and-a-half Belgian francs. (At that time, the Belgian franc was valued at twenty-five francs to the pound sterling.) The Tordus barrel was but one of a considerable range of patterns, some very intricate, made in the Vesdre Valley and exported to the gunmakers of the rest of Europe and North America.

It is because of their distinctive and often beautiful patterns that we are able to identify the original source of these barrels and so get a glimpse of the international trade that existed in gun parts as well as in finished guns.

From the point of view of an artisan working in the gun trade in Birmingham, this movement of parts was a cruel blow. Apart from studying the barrels, there was no way of learning the origin of the material. Cheap imported parts could be assembled in Birmingham and, when they had acquired the marks of the Birmingham proof house, sold as English guns. In this way, the reputation of the English gun failed to provide work for the English gunmaker. Since British politicians of all parties were unable or unwilling to curb this trade, the long-term effect was the decline of work in the Birmingham gun trade.

Gilbertson & Page of Watford. Game food advertisement from *The Shooting Times*, 1898

Above and below: From *The Shooting Times* Christmas issue of 1898

"THAT WAS A HIT GILES. DIDN'T YOU SEE THE FEATHERS FLY?"
"YEZZUR - BUT THE BURD FLEW WI'EM"

"A RIGHT AND LEFT AT WOODCOCK."

CHAPTER SIX
A Gun for a Poacher

For a poacher, buying a new gun would have been an option of the very last resort. The vagaries of the profession were such that he, or she, would have regarded such an outlay as a very risky investment indeed.

In view of this, our poacher's attention would have been attracted to the very bottom of the breech-loading market and here he would have found guns as remarkable in their way as any on offer, with both more diversity of design and a far wider range of origins than anywhere else.

At the bottom of the price scale was a selection of single-barrel guns that had begun their careers serving with one of the huge armies of continental Europe. It should be explained that, in the course of the rapid technological evolution and its attendant small arms race that characterised the second half of the nineteenth century, vast numbers of muzzle-loading muskets and rifles had been made redundant. As the rush of innovation continued, they were joined by large calibre, single-shot rifles. Some of these gravitated unaltered to second-line troops at home and to warlords and bandits of all sorts in various far-flung locations, but their outlets could not cope with the sheer volume of obsolete arms. The residue seems to have found its way to Liege at prices which must have been only a very little above their scrap value, for how else could the resulting conversions have reappeared for retail sale at just a few shillings?

Inevitably, the minimum of work was done to enable a shotgun cartridge to be fired. So much of the appearance of the original arm remained, certainly the shape of the stock and the balance and handling characteristics of the military

12-bore shotgun converted from a Chasspot rifle

HEYDAY OF THE SHOTGUN

The *Tabatière* alias The Zulu. Breech closed: note the firing pin is missing from this specimen

The *Tabatière* alias The Zulu. Breech open to show details of the extractor

weapon were retained. Thus, while the remodelled piece could be described as a shotgun, the term 'sporting gun' would have been dubious. However, it could be argued that the very qualities that made the original weapon recruit-proof in an army composed of peasant conscripts, would have still been of value to its new owner.

Obviously the original configuration of the arm determined the work needed to be done to adapt a musket or rifle to its new role. For the muzzle-to-breech-loader transformation the most usual method was that colloquially known as the '*Tabatière*' or snuff box because of its supposed resemblance to such a container. This was the design of a man called Clairville. To English eyes it is a crude version of the Snider and indeed some writers have gone so far as to describe it as an inferior copy. However, this is not the case and in terms of patent dates, the Clairville predates the Snider by some nine years. Moreover, it differs in being shorter than the Snider breech block and because of this, to accommodate a cartridge of the proportions of a 12-bore, a cutaway portion has to be created in the head of the stock.

While the specification for the *Tabatière* shows pinfire civilian pistols, the first widespread use was as a means of converting French military muzzle-loading rifles to breech-loading. Not surprisingly, the same system was employed to convert both complete muzzle-loaders and assemblages of cannibalised parts to centre fire shotguns. Less economically credible is the claim that totally new guns were made in this fashion. For some reason, the *Tabatière* was known in the United States as a 'Zulu' and possibly this was originally a brand name.

Without the benefit of a catchy name, the remodelled military arm that is most often encountered in Great Britain is the bolt-action, single-shot that started life as the French needle-fire Chasspot rifle, best known as the first-line weapon of the French army in its disastrous war with Prussia in 1870 and 1871. The Chasspot is named after its inventor Antoine Alphonse Chasspot, who, from 1864, was principal of the arms factory at Châtellerault, north of Poitiers in western France.

Unlike later bolt-actions, opening and cocking are two separate motions on a Chasspot. One feature of this action that was of vital importance to all its users, especially new civilian owners, was that there was no safety catch or half-cock position. The bolt is merely held in the cocked position by a sear engaging with a bent on its underside.

There can be no doubt that some of the converted rifles that appeared on the British market for about seventeen shillings (85p) had seen active service and, therefore, were worn before they were altered. They were thus on the way to becoming dangerous even when they were bought by their new owners. As a result of wear to the bent and sear, a situation was created in which any jolt was sufficient to fire the gun. When this alarming and potentially fatal characteristic was appreciated, the lucky owner would always be accorded the privilege of leading a file, because nobody dared walk in front of him.

This lethal reputation, the awkward military stock and the extra length conferred on the gun by its bolt-action counted seriously against the Chasspot conversion. However, what had been demonstrated was that there was a definite call for a cheap, breech-loading, single-barrel shotgun on the British market. So, when the supply of converted Chasspots eventually dried up, the importers cast around for something else to take its place.

They found that a type of gun which came very close to filling their needs was one of the staple items of a number of makers in the north-eastern corner of the United States. The cheap, drop-down, single-barrel shotguns that were being produced owed their origins to the existence of factories that had flourished during the period of the Civil War, but had become redundant overnight when peace returned. It was a relatively small change from making single-barrel, single-shot rifles for the Federal Army to making single-barrel shotguns for the new army of settlers who moved west at the end of the war. Indeed, at the beginning, redundant barrels intended for rifles of about .58 calibre were bored out to become 12-bore shotgun barrels, there being ample barrel wall thickness for such an alteration.

The only real snag with the American guns, as far as the British market was concerned, was their price. Being new and even with the advantages of

Above: Harrington & Richardson, USA. Machine-made single barrel 12-bore

Right: Harrington & Richardson, USA. Logo

the water power from the Adirondacks to drive the machinery that made them, they could not match the price of a converted Chasspot. By the time these Yankee guns had been shipped across the North Atlantic and been English proved, their retail price had climbed to between thirty shillings (£1·50) and two guineas (£2·10). At the top end of this range they cost more than double the price of a Chasspot. This increase was more than the gross weekly wage of a farm labourer at the time.

Naturally the vendors made much play of the fact that these guns were English proved, conveniently leaving aside the fact that British law required this. A curious anomaly existed in that, despite being a major arms producer, the United States did not have an official proof house and therefore no test that could be recognised in this country.

For some reason, only a relatively small selection of the American makes were imported into the United Kingdom. The firm that discovered America as a source of such guns seems to have

J. Stevens Arms and Tool Company.

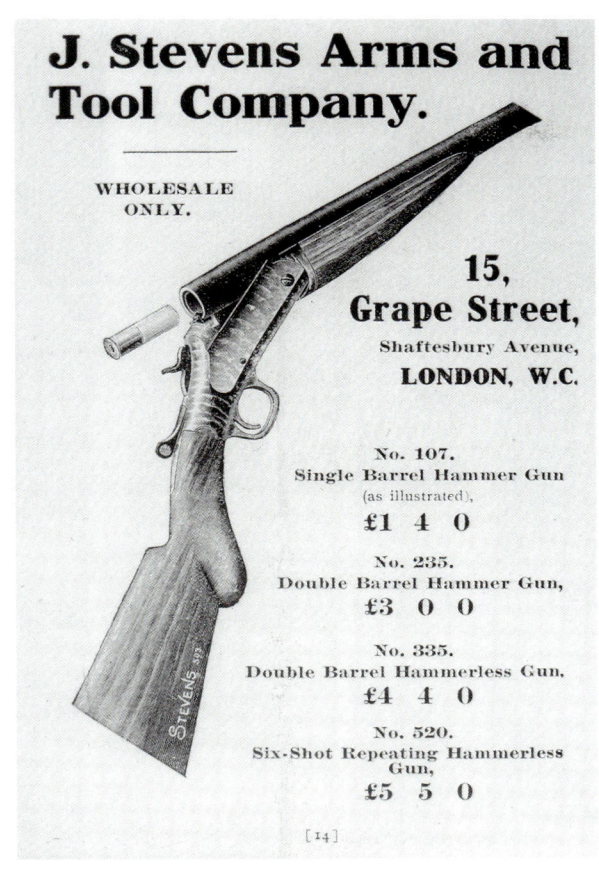

WHOLESALE ONLY.

15, Grape Street,
Shaftesbury Avenue,
LONDON, W.C.

No. 107.
Single Barrel Hammer Gun
(as illustrated),
£1 4 0

No. 235.
Double Barrel Hammer Gun,
£3 0 0

No. 335.
Double Barrel Hammerless Gun,
£4 4 0

No. 520.
Six-Shot Repeating Hammerless Gun,
£5 5 0

[14]

J. Stevens Arms & Tool Company advertisement, c. 1910

Unnamed Belgian hammer gun, 12-bore top lever

---

because the Harrington & Richardson fast became the single by which all others were judged. This is not to imply that it was the only American single being imported. Remington, Ivor Johnson and Stevens guns also appeared, but in much smaller numbers than the 'Harringtons'.

If we assume that our poacher had a good season and fancied a real double-barrel gun rather than the converted military arm or the curiously stocked Yankee single, his eye might well have lit on the temptingly-priced guns sold by the ironmonger or cut-price gunmaker. These guns, all made in Belgium despite attempts by the use of English-sounding brand names to imply an alternative origin, were the staple of the bottom of the British new double gun market, being priced at about £2. That it was possible to retail them at this price, having generated at least two and possibly more intermediate profits on the way, calls for an explanation of how these guns were made.

In a world that was moving towards the concept of ever larger and more mechanised factories, the gunmakers in the villages around Liege in eastern Belgium represented an anachronism. They were, in every sense of the term, a cottage industry. What the gun work represented was a cash crop, a means by which a little money could be earned to buy those few necessities and minor luxuries that a small-holding with its vegetable plot, orchard, cow, pig and a few chickens did not provide. Central to understanding this economic system is the fact that each man was his own master. Assuredly he may have been poor, but to compensate he had the treasure that is personal freedom – the freedom to decide when he worked and to have his wife and children to help him and learn the trade by a natural apprenticeship. In essence, the Liegeois gunmaker was no different to the out-worker in Birmingham but, with the land providing most of his basic needs, he could, and did, accept lower rewards for his labours.

Another remarkable feature of the cheapest Belgian guns was that they were largely conventional in design. So often with all sorts of manufactured items, the search for a low final price starts with a redesign and a radical simplification of the product. Interestingly, at the very same time as the cheap Belgian guns were being filed out in rural workshops, just such simple designs were being

been the Birmingham-based gun-making entrepreneurs trading under the name of Charles Osborne from Whittall Street. In 1901 they began to import the Harrington & Richardson single-barrel and thereby made the wisest choice,

made in the gun factories of New England. For instance, the Baker Gun Company of Batavia in New York State made a lock of brilliant simplicity.

Logic would suggest that, if a mechanism like this were taken up by the Belgians, an even cheaper product would have resulted, or perhaps more profit for the middle man. Instead, a surprising proportion of the Belgian guns retained a design essentially the same as the very best gun made in London. Speaking generally, there were only two concessions to simplification on the Belgian gun.

The usual lock has what is termed a two-pin bridle, which means that the inner plate or 'bridle' that provides the second bearing for the tumbler and sear is held in place by two screws or pins. The best English locks had come to have four or even five pins and, it was claimed, a more stable and solid bridle as a result. This was probably an unnecessary refinement. After all, in the 1850s and 1860s, the pinfires and early centre fires produced by the best makers used locks with three-pin bridles.

The other noticeable design feature of the Belgian gun was the use of barrels that were round for their entire length, that is to say there were no flats under the breech ends where they lay on the bar of the action. Instead, two trough-like depressions were cut fore and aft on the action bar. The same mode of construction is to be found not only on converted muzzle-loaders, but also on guns of the first quality so, like the simple lock, this style of construction cannot be regarded as the sole preserve of the cheap gun.

Instead, to achieve the ridiculously low price, both the quality of the materials and the time spent by the workmen had to be reduced to an absolute minimum. Wood was used for stocks that the maker of fine guns would consider suitable only for firewood. Locks and other moving parts were filed from soft iron instead of hardened steel and so wore rapidly to produce in short order a gun that rattled when shaken. The wear in the lockwork produced a

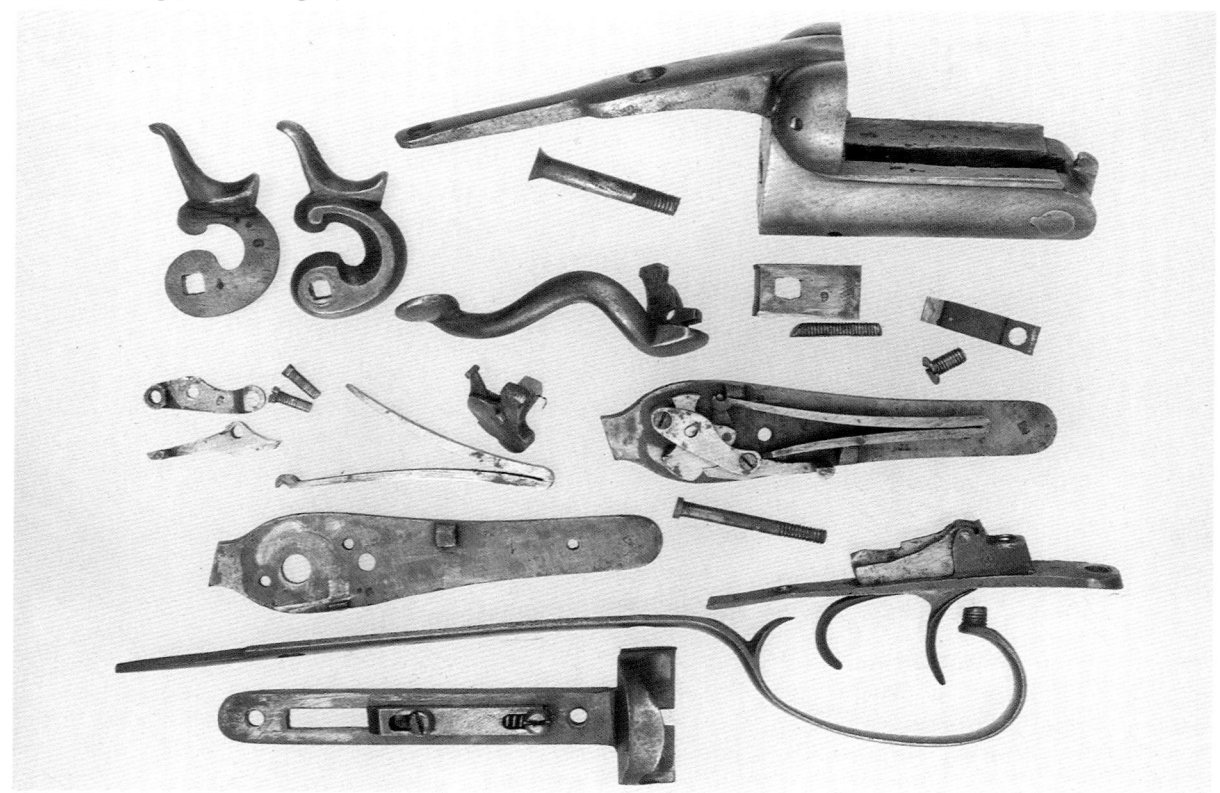

Unnamed Belgian hammer gun, 16-bore side lever

dangerous situation when the trigger pressure could either become so light that the lock fired as if it had a 'hair trigger', or conversely became so heavy that the proverbial traction engine was required to pull the trigger. Perversely, guns are found in which one lock has gone light, the other hard, so the differential is even more accentuated. It is, of course, not necessary even to take off a lock to appreciate the quality of the workmanship. A look towards the light down the outside of these cheap guns to see how barrels should not be filed up is enough.

The catalogue of defects covers every part of the gun, literally from muzzle to heel plate. The remarkable fact is that these crude guns did not fail more often in use. There are two possible explanations for this. Firstly, in comparison with a sportsman on a large shoot, the user of an ironmonger's gun would have fired very few cartridges. The differential was such that, on a big day, a 'toff' might well have fired enough cartridges to keep a pot-hunting rustic and his family in rabbit dinners for a decade or more. Secondly, the cheapest cartridges generated lower breech pressures because they were loaded with black powder. They retailed at a price of about six shillings (30p) a hundred, not that the likely owner of such a gun would have bought a hundred cartridges all at once. To cater for such customers, small-town ironmongers sold cartridges individually.

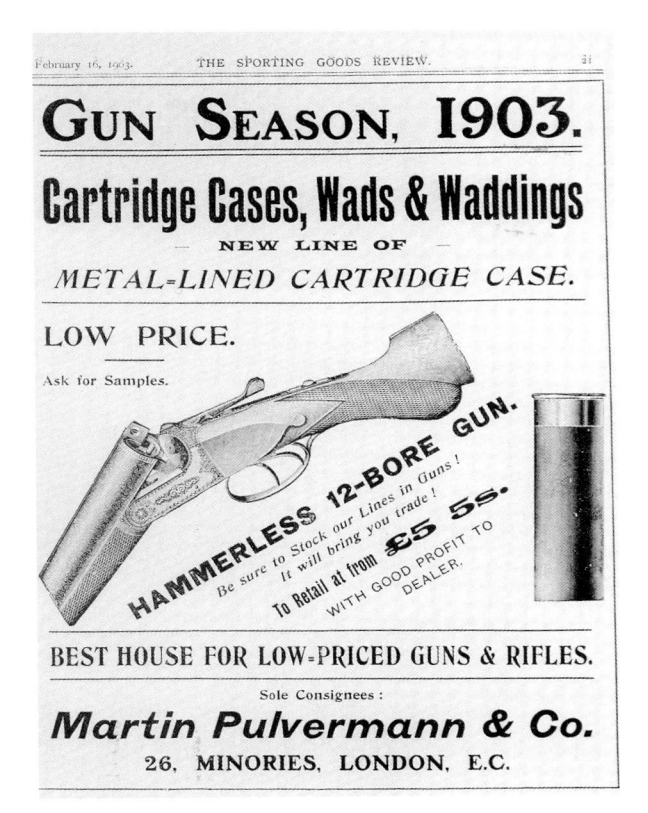

However, with the constraints of a finely-balanced budget and a lack of any technical knowledge, the Belgian proof marks would almost certainly have been a total mystery to the owner. The culture would have been to fire any cartridge that could be crammed into the chamber. Given that these cartridges could vary from the best pilfered on a smart shoot, to cheap ones loaded by the ironmonger's boy on a wet afternoon, via all sorts of dubious home loads, damp rounds cooked in front of the fire with the odd long-cased load thrown in for good measure, the Belgian 'rattle trap' could have been faced with a very mixed diet.

Against such a background, the conventional wisdom is that these guns should have failed with burst barrels or cracked actions, that they did not only goes to prove that there must really be a guardian angel who looks after fools.

Martin Pulvermann & Co. advertisement in *The Sporting Goods Review for 1903*

CHAPTER SEVEN
A Gun for a Conservative

There are many reasons why people resist change but, in the case of their sporting gun, the cost involved acts as a powerful incentive. Given the fact that a gun of middle or better quality, if properly looked after, will serve at least two shooting lifetimes, a son can inherit his father's guns and need no others.

In 1900 gunpowder could be bought easily and readily and could even be sold to children with no more formality than a pound of tea. Therefore, there was no bar to the use of the muzzle-loading gun. True, muzzle-loaders had been taken to the very fringes of sport, in the shape of punt guns and large calibre shoulder wild-fowling guns.

In fact, in rural England, the muzzle-loader was used most commonly by the village boy sent to 'starve the crows' by keeping them off the crops. Officially, he would be issued with a supply of

---

A gun for a crow starver

REMARKABLE GUN ACCIDENT.

A REMARKABLE accident is (writes a correspondent) under treatment at Spalding Johnson Hospital. A farm servant named Arthur Doades, aged fifteen, was in a field with a gun scaring birds, and when reloading the weapon and ramming down the charge, the gun exploded and the ramrod passed through the boy's head. It entered his forehead, and came out at the top of the head, carrying his cap with it, alighting some distance away. He was found in an unconscious state, and at first no hopes were entertained of his recovery; but though the injury to the brain is serious, he has now recovered consciousness, and there is a slight chance that he may get better. The case is regarded at the hospital as a most remarkable one, and the fact of his still being alive is attributed to his exceptionally robust constitution.

A gruesome reminder of the dangers of muzzle-loading, the unfortunate lad is believed to have survived

powder, probably some newspaper for wadding and a few caps if it were a percussion ignition gun. However, boys will be boys and a supply of shot acquired in some way or even a few judiciously selected stones would serve to bag an unwary rabbit or other game, which would have been very welcome on some cottage table.

Ironically, while an urchin from the village might be up to such mischief, his privileged counterpart from the big house might also be shooting with a muzzle-loader. For it was believed by some conservative mentors that a boy was best taught to shoot with a muzzle-loader, in the same sort of way as sailors are trained in sailing boats. The art of shotgun shooting with a muzzle-loader was made easier for the raw beginner, since it was possible to vary the charge infinitely, so allowing him to start on a gun that was practically without recoil. Then, as his technique improved, the charge could be judiciously increased to that commensurate with the weight of the gun.

The main argument advanced for the use of the muzzle-loader as a gun for a learner was that he would not blaze away, but learn to pick his shots. Unfortunately, picking shots can mean dithering and it is debatable which is the greater vice.

For such training purposes, a single-barrel gun is preferable, even essential, because with a single all the potential problems of muzzle-loading are reduced to a minimum. It is still possible to double charge, not ram wads tight enough, or to forget to put in either shot or powder. In the same way it is still possible to leave some cleaning material in the barrel, which remains glowing from the first shot and ignites the second charge of powder along with the whole powder flask. However, the presence of an old hand should take care of such problems, while the single barrel eliminates the danger of dealing with the muzzle of a gun with one barrel still loaded.

It could be argued that, with a muzzle-loader, the learner could not be taught the etiquette of shooting in company with a breech-loader. For this reason it was desirable that the muzzler be

W. Brummitt of Mansfield muzzle loading single barrel gun approx 28-bore

replaced as soon as possible. However, the advocates of ritual training with the muzzle-loader could point to the fact that this was the way in which the Duke of York (the future King George V), and a game shooter with above average ability, had been taught.

Before we leave the subject of the muzzle-loader, we must note that by choosing such a gun the ultra conservative skinflint could buy the most beautiful double-barrel gun for a tenth of its original cost. There was minimal interest in such items from collectors who were more inclined towards eighteenth century guns, or the best flintlocks of the early nineteenth century. True, a percussion muzzle-loader and its owner would have been the butts of all sorts of derision if the gun were carried on a formal shoot, but if a man wished to potter about on his own or with a friend, and reduce the costs of his ammunition to an absolute minimum, he could shoot with the very best that London could produce.

For most of the conservative clan, the choice would have been a breech-loader. For many, in terms of quality, a pinfire offered the best bargain. Pinfire cartridges were still freely available, although not perhaps in the bewildering variety that existed with centre fires. For instance, the 1902/3 Kynoch catalogue lists no less than eleven

Detail of the escutcheon of W. Brummitt gun

Below: C. Guest, place of business unknown: possibly the original owner. A double-barrel muzzle-loader of superb quality, the style is very reminiscent of the work of W. Greener

different 12-bore centre fire cartridge cases, with different constructions, not merely length variations, but there were only three designs of pinfire. Not that this could be called a serious handicap.

Much can be made of the inconvenience of using a pinfire, the need to locate the pin accurately in the slot in the top of the breech or the discomfort of the pins digging into the flesh when carried in a pocket. There is also the story that they would explode if dropped. As with all theoretical claims, practice isn't quite the same thing. For instance, it is perfectly possible to reload a pinfire gun in the dark or without looking at it at all. The trick is to position the cartridge roughly and, with barrel muzzle down, simply rotate the cartridge until the pin drops into the slot. There is no problem of feeling when it is in its proper place.

The pins are painful if carried in a trouser pocket and the answer is a cartridge belt with all the pins lying the same way, like the spines of a hedgehog. After all, you can stroke a hedgehog.

I am unaware of any documented proof of accidental explosions, although this is a frequently repeated claim. Certainly, when dropped from waist height, the tendency of a loaded cartridge is to fall heavy end first, in other words shot down.

In use with cartridges of acceptable quality, there is always a certain degree of gas escape round the pin and the gun is a little slower to reload. However, in this context, we should not lose sight of the fact that in August 1872, Sir Frederick Milbank had shot ninety-five brace of grouse in twenty-three minutes using a trio of Westley Richards pinfires.

A rather different type of conservative was the shooter who clung to the single-barrel gun. In so doing he was heir to an ancient tradition and the origins of shooting with sporting shotguns. For it was, after all, only about the end of the eighteenth century when the double-barrel sporting gun became practical. For about 150 years the long, single-barrel flintlock fowling piece had been the only gun. When the double gun appeared with the option of discharging two charges at one bird, there were those who said that the new gun was unsporting. The view was that when your partridge evaded your shot, if you were a sportsman you would concede your defeat gracefully. The evidence is that at the beginning of the nineteenth century such views, and it must be added the dangers inherent in a double-barrel muzzle-loader, were sufficient to induce a good

W. Evans, London. Single-barrel 12-bore

proportion of sportsmen to adhere to the single-barrel.

As the century progressed, and certainly after the breech-loader made the double-barrel gun infinitely safer, the single-barrel waned in popularity, but did not entirely vanish. On a practical level, because such a gun is less bulky and has a weight distribution that tends to the stock end, a single somehow feels less of an encumbrance. Due to these attributes, it became the boy's gun or the lady's gun. A sporting landowner out to survey his property not specifically to shoot, but who might wish to bag the odd brace for the pot or deal with a crow, would also carry a single on his rounds.

Certainly, gunmakers' catalogues, and even more persuasively surviving specimens, indicate that as late as 1900 enough of the ways of the seventeenth century persisted to sustain a limited trade in the high quality single-barrel, single shot gun.

Only slightly less venerable than the single was the converted gun, another preserve of the conservative. All down the ages existing guns and rifles had been altered to take advantage of innovations. Matchlocks had been converted to various mechanical forms of ignition and, in some cases, gave another century's service thereafter.

In earliest times the motivation lay, in part, in the intrinsic value of the barrel. A fine barrel was greatly prized. So we find barrels fashioned from the steel of the east with its characteristic 'watered' damascus pattern, captured after the rout of the Turks at Vienna in 1683, reused on western guns. At a later date, British travellers to Spain would bring back a product of the famed barrel smiths of that country and use it as the basis of a gun built to their order by one of the premier gunmakers in London.

The nineteenth century had started with sportsmen using flintlock muzzle-loaders and it was to become the century of the gun converter. The first half of the century saw the wholesale transformation of flintlocks to the new percussion ignition. Inevitably, the quality of this work varied.

Joseph Wilkes, Leeds. His trade label offering conversion of muzzle-loaders to breech-loaders

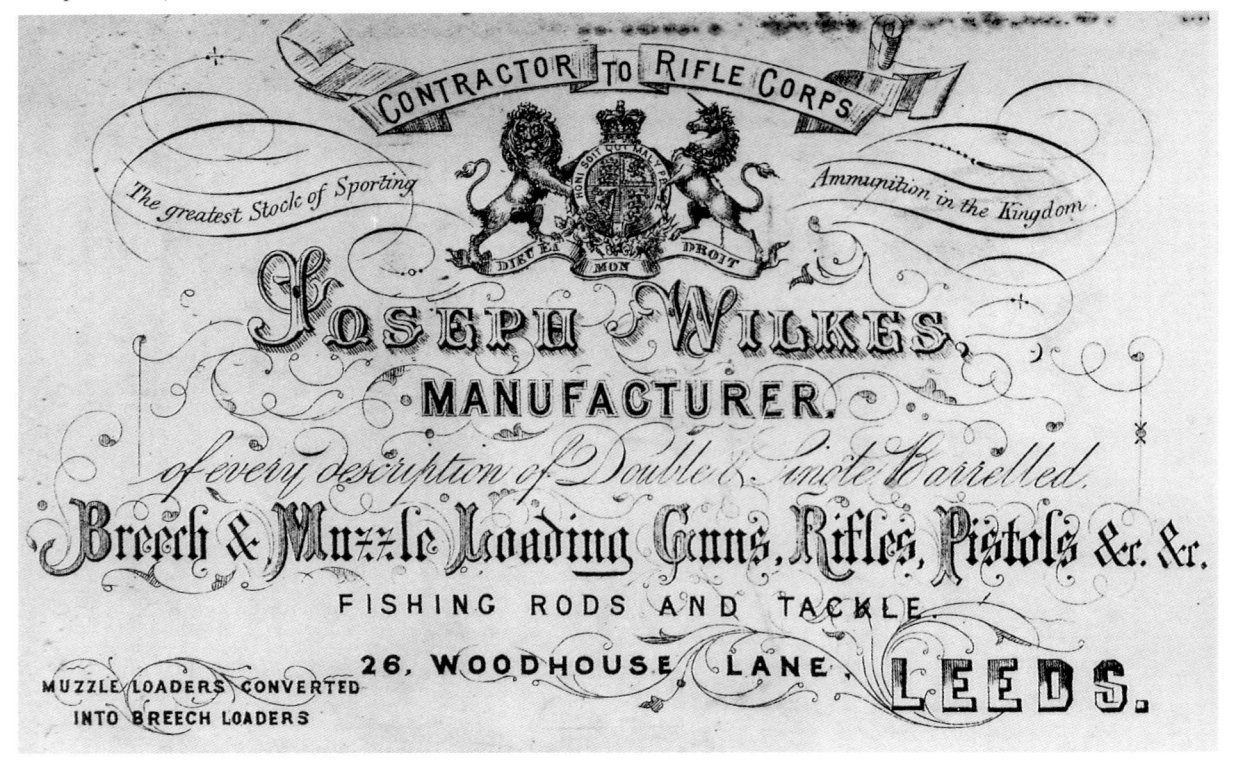

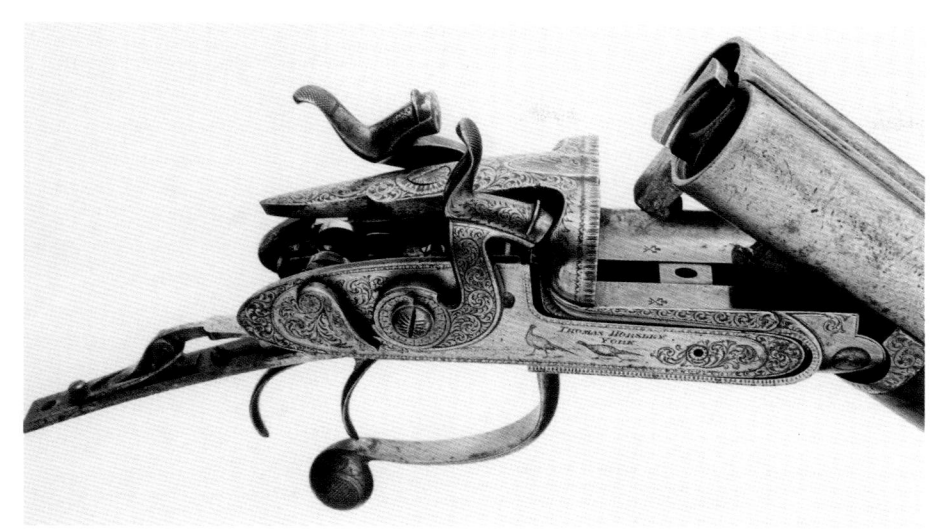

Thomas Horsley, York. A 12-bore muzzle-loader converted first to pin thence to centre fire

At best a gun would be rebuilt by its original maker and, if the gun was relatively new when rebuilt, styles would not have evolved. It is, therefore, all but impossible to guess the history of the gun concerned. At a lower, but still acceptable, level we see neat bodges, good for indefinite service. In truth, the same could be said of the crude work which characterised the very bottom of the market.

From our present perspective, the importance of all this activity, especially the flurry of conversions from flint to percussion, was that it served to set the scene for the arrival of the breech-loader. The transformation of a muzzle-loader into a breech-loader obviously required major surgery, far outclassing anything that had gone before. By comparison, the simplest of conversions from flint to percussion was mere tinkering. Amazingly, it was carried out routinely and very successfully

In a sense this fashion had been sanctioned, if not actually set, by the government with the conversion of something like half a million Enfield muzzle-loading rifles and carbines to the Snider system breech-loader. Conversions on a similar scale had been undertaken by most of the major armies of the western world, but it would have been with the Snider in the Volunteers that most British sportsmen first encountered a breech-loader rebuilt from a muzzle-loader. It is important to recognise the influence of the Militia in late nineteenth century Great Britain, indeed some authors went as far as to classify 'Volunteering' as a field sport.

The Snider demonstrated on a vast scale that, properly done, a converted muzzle-loader was a perfectly sound, practical option. Naturally, the government had the advantage that the Enfield rifle was a machine-made, standard item, which is why one could be converted for less than a pound, one of the stipulations issued by the government at the beginning of the exercise.

If we are to make sense of the economics of gun conversion at the end of the nineteenth century, we must realise that this price represented something like half a week's wages for the men engaged at Enfield Lock, who worked a fifty-nine-hour week for eight pence (3p) an hour. Prices of the order of £15 quoted for a top-class job begin to come into focus. As always, there was a variety of options and in *How to Buy a Gun*, published in 1903, prices range from a scarcely credible £3 up to a more reasonable £10.

However, before we get embroiled in the actual work involved we have to consider what characteristics made a muzzle-loader suitable for conversion. The actual calibre of a shotgun was less critical than that of a rifle, unless, of course, the extra labour of re-boring the rifle barrel was to be undertaken. What was vital was that there should be sufficient thickness in the barrel walls at the breech to permit the cutting of an appropriately-sized chamber.

This problem was compounded by the fact that

the outside profile of the barrel of a muzzle-loader is more of a straight taper from breech to muzzle. There is none of the flair at the breech end that is characteristic of the breech-loader. The inevitable result is that the chamber walls of a conversion are noticeably thin, indeed this is often the first clue to spotting a conversion. Because these barrels survived reproof, they have to be accepted as being adequately strong. The greater problem of thin chamber walls was that they made the drilling of holes for the extractor leg highly critical. Indeed, with the advent of smokeless powder, compromised chamber walls were a fruitful source of bursts.

Assuming that the barrels were suitable, the customer would need to choose the barrel bolting action for his 'new' gun. While there were patented designs claiming to be especially suitable for a conversion, in fact almost any action could be used. Reflecting its wide use on guns, and indeed rifles of all sorts, many guns are found with the classic screw grip 'Henry Jones' action. For some reason the forward-facing side lever snap Smith patent of 1863 is widely seen. Beyond that it is impossible to be dogmatic because so many different actions are encountered in small numbers.

Whatever the bolting system, there was a need to fit lumps to the underside of the barrels. Two methods were employed. In the simpler version a piece of steel was fitted and brazed into the 'V' shape between the two barrels and the new action bar was formed with channels to receive the curved undersides of the barrels. In the other a 'saddle lump' was fitted right across the bottom of the breech portion of the barrels. This had the advantage of producing both a greater area of contact to braze lumps to barrels and could also be shaped to true action flats.

Whichever action was chosen, the gunsmith faced the problem of fitting it to the existing stock. Skilled men stand in awe of the expertise that this action filing represented, because it is the reverse of the usual practice where wood is fitted to metal.

The earliest conversions from muzzle-loader were those to pinfire ignition cartridge breech-loaders. Some of these guns were then converted

A. Agnew, Welshpool. 12-bore, Agnew's patent action (1863). Amateur conversion pin to centre fire

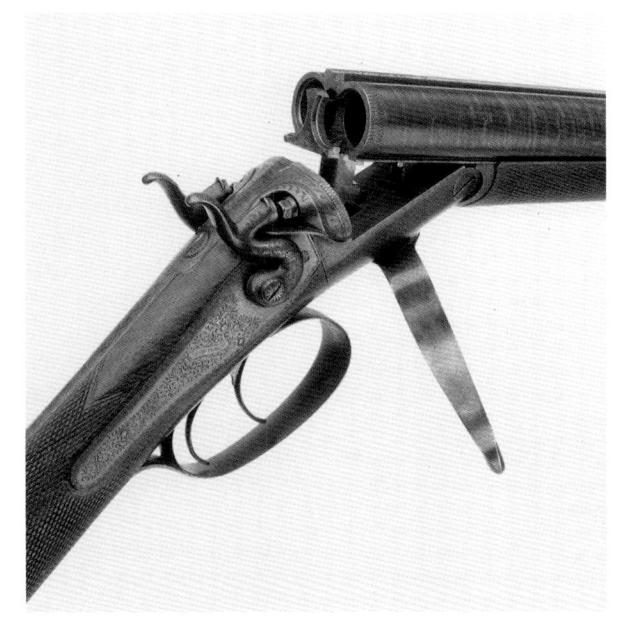

J. Squires of London, 14-bore centre fire, back-action locks, screw grip Henry Jones action. A very skilful conversion of a pinfire to centre fire

again to take the centre fire cartridge or, alternatively, guns built as pinfires were converted to centre fire. So, before we leave the question of actions, the special problems of the pin to centre fire conversions need to be considered.

The essential extra needs of a centre fire cartridge when compared with a pinfire are a firing pin, a bigger recess for the cartridge rim and an extractor to act on this rim. The most basic conversions to centre from pin added only these features and made do with butchered pinfire cocks to strike the new pins. On a slightly more sophisticated level, new hammers were fitted and, while the result was still an obvious bodge, it served its purpose as a provider of rabbit dinners.

The best conversions were rebuilds on a par with the best flint-to-percussion transformations. Many of these guns have been so cleverly altered that there are plenty of examples of guns being owned for a shooting lifetime without this aspect being appreciated by their owners. The skill of the converter lay in altering the slope of the standing breech and not in simply reshaping the existing thin, almost flimsy, pinfire component but adding metal and filing up a totally new form. The extra metal could be added at two sites. The first and more ingenious, was at the breech face where a plate about three millimetres thick was added. Not only did this add the necessary bulk to the standing breech, but it enabled the same amount to be lopped off the breech end of the barrels, thereby eliminating the pin slots.

For want of a better name, we can call these additions 'breech plates'. At first glance from the outside they look like an attractive band of engraving on the front of the standing breech. The engraving makes a convenient camouflage for the joint. The presence of a breech plate can usually be detected by looking at the breech face, especially on a worn or one-time corroded gun.

J. Squires, London. Top view to show addition to standing breech

Here will be seen four faint circles, one either side of the firing pins and two, one above the other, on the centre line of the gun. These are the studs which are screwed into the original standing breech and then struck off flush with the new breech face.

The other usual way to add metal to the standing breech is to fit inserts on either side of the top strap in the angle where this joins the standing breech. Sometimes these additions are betrayed by a 'gold' line of braze visible on the top of the new breech. Often, however, all that can be seen is a very faint line if the light is made to fall in just the right way across the breech. With this extra metal at his disposal, the action filer can create a totally new shape to the action and so alter the style of the gun completely.

Inevitably, the progression of the converted gun shadowed the evolution of the sporting gun and so the rebound lock produced a new and lucrative opportunity for the conversion specialist. Since this change involved only internal parts of the gun, the change can be almost impossible to detect if these new parts match the old in terms of quality of finish. However, if the style of the gun suggests that it is an early centre fire, then conversion should always be suspected, because it was such a useful modification.

In many respects the conversion to rebound marks the end of an era. This is not to say the cult of converting ceased, in say, 1880. The conversions described continued to be performed, indeed the pin to centre fire was a means of employing men gainfully in slack times. However, the hammerless gun with choke-bored barrels seems to have induced men to buy new, not that conversion to hammerless was not advertised and indeed carried out, but on a much reduced scale compared with muzzle-to-breech-loader.

The problem essentially is one of layout. A hammer gun tends to be narrower at the head of the stock than a hammerless one. Moreover, fitting suitable lockwork is further complicated by the fact that the standing breech of a hammer gun is higher in relation to the stock than is normal for a hammerless. On this point it is worth noting that Thomas Perkes, in his advertisement, specifically mentions a pinfire to hammerless conversion, where the problem of the height of the head of the stock would be less critical. The problem of the narrowness of the head of the stock would remain, so the firing pins would have to splay outwards.

By comparison, the conversion of a non-ejector gun, be it hammer or hammerless, to ejector is a

Rebound and non-rebound locks

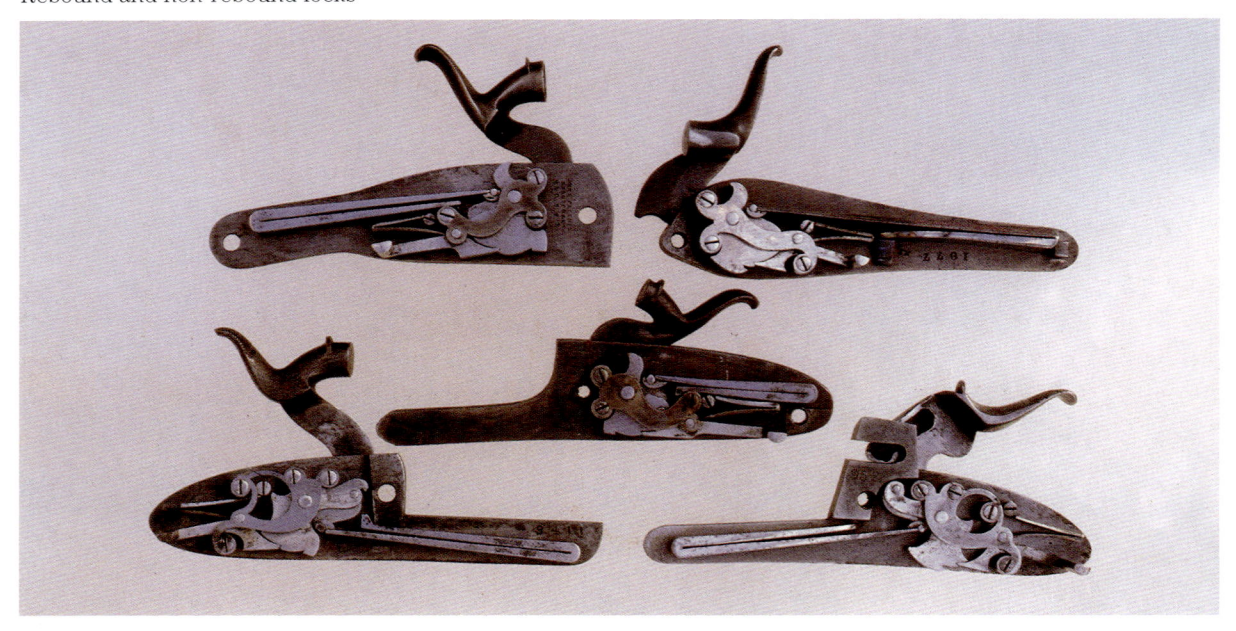

Group of unknown shooters. Their dress suggests *c.* 1900. Two bottles of beer will not go far!

simple task. There are of the order of a hundred different ejector patents, so the choice of mechanism is wide. Most of these are built into the forend of the gun and simply require some sort of trip rod activated by the lockwork to bring them into action.

The single trigger also offers a huge choice of mechanisms with the advantage that an existing component is removed to make room for it. Here we encounter a strange phenomenon in that guns in use are converted back and forth in both directions. Guns built as 'one-triggers', to give them their nineteenth century term, are altered to double triggers and vice versa.

The external hammer gun had reached the pinnacle of its development by the mid-1870s and so by 1900 could be regarded as a proven standard item. The two design features that may be said to complete the perfected gun are the rebound lock and the double snap bite.

A rebound lock is a mechanism in which the hammer only rests in two positions, the full cock and the half cock. The internal works are such that, immediately after striking the firing pin, the hammer bounces back to the half cock position. The lock is, therefore, safe and the gun is ready to open the instant that the gun has been fired. These two great advantages meant that the rebound lock was quickly adopted across the whole spectrum of makers and qualities.

Not quite the same universal approbation was offered to the double snap bite action. True, it was widely used, but not to the exclusion of the previous industry standard, the turning underlever working a double-bolt, screw grip. To illustrate this point, where the guns can be identified in the group of shooters pictured, all are hammer guns. Of these, two and possibly three, are screw grip underlevers.

While most of the conservative shooters so far considered were, in part at least, motivated by the cost of a new gun, there was one group who were not. This was a coterie of some of the best shots in Great Britain, the owners of thousands of acres and fine mansions who continued to shoot with and order new hammer guns. Given their prowess on the shooting field, the names of the Duke of York and Lord Walsingham are well known in the context. They were not alone. Makers' records and surviving guns tell us that other, less well-known game shots followed this persuasive example. The guns they bought were not merely ejector versions of those produced in the late 1860s when the hammer gun was first perfected. The hammer ejectors of the 1890s were designed to consume without complaint tens of thousands of nitro cartridges. So we see the use of slightly heavier actions. Purdeys reverted to back-action locks to ensure that there was plenty of metal in their action bodies. Many of these late hammer guns were fitted with steel barrels, but based on his very extensive experience Lord Walsingham had a change of heart, and had Purdey's rebarrel his famous trio of hammer guns with damascus tubes.

However, the most noticeable feature of the late hammer gun is the smaller size of its hammers. This is strange, given the claim that the hammers were some sort of sighting aid. This oft-made claim is difficult to evaluate but, on balance, given that the trap shooters with money and with their prestige depending on every shot, were not faithful to the hammer gun, the balance of probability lies in favour of conservatism being the true explanation.

Whatever the reason for conservative foibles, they do have a certain charm. Certainly the shooting field would have been the poorer without them.

Boss & Co., London. 12-bore top lever hammer gun with bar-action locks

CHAPTER EIGHT
A Gun for a Trap Shooter

The story of live pigeon trap shooting in Great Britain is one of swings of fashion. The pastime is first recorded at the end of the eighteenth century, when an article in the *Sporting Magazine* claims that it was practised by 'the first shots in England'. However, this popularity was soon to wane and it became 'a recreation of the country pot house'.

The 1820s saw another high spot in the story when trap shooting became one of the mediums for the reckless gambling that was a feature of this age. There was, for instance, the famous match in June 1827 between Lord Kennedy and 'Squire' Osbaldeston, in which they each shot at one hundred pigeons a day for four days, for a bet of 2000 guineas a side. Again the wheel of fortune turned and by the 1840s the pastime had reverted to the inn yard.

In the middle of the nineteenth century, a group of aristocratic shooters revived trap shooting at Hornsey Wood House which was situated in what was then the outskirts of London but which is now the north-west corner of Finsbury Park and the adjacent roads. From the evidence of surviving prize guns, it seems that they called their club or organisation 'The 100 Gentlemen'. Unfortunately, there is no known documentary record of this club, even though some of those recorded on the prize guns were later subjects of biographical works. Involved in some way with the club was a Frank Heathcote, some accounts call him the manager, some the organiser, but most credit him with the introduction of a system of handicapping and framing rules to exclude large bore guns. The shooting at Hornsey Wood flourished until about 1867 when the ground ceased to be available,

The crowd watching a shot. Sidney Smith. (PHOTOGRAPH COURTESY OF THE YORKSHIRE MUSEUM)

whereupon Frank Heathcote rented the Hurlingham Estate at Fulham and The Hurlingham Pigeon Shooting Club was formed in 1868. Almost from the beginning of its existence, The Hurlingham Club became the premier venue for live pigeon shooting in Great Britain. Men of fashion from the Prince of Wales downwards shot there and for several years there were annual matches between teams drawn from the two Houses of Parliament.

The inevitable result of this endorsement by the highest in the land was that live pigeon shooting clubs sprang up in most of the principal cities and spas, not just in Great Britain but across Europe, North America and on to the Antipodes as well. As a natural consequence, competitive live bird shooting also spread down the social scale where it appealed not only to those participants attracted by the tempting prizes but also to men debarred by social position from any other legitimate shotgun shooting. At its lower levels and to reduce costs, small wild birds were captured and released as marks, for instance, house sparrows and starlings in Great Britain and parakeets in Australia.

One thing this activity was good for was trade, all the way from the most prestigious gunmakers in the West End of London who sold guns to the gentry to use at Hurlingham, right down to the urchins who caught sparrows for the match behind the less-than-salubrious public house. It could, with truth, be claimed, that each pigeon generated several profits, first for the small farmer in Lincolnshire who raised it, for the dealer who sold it to the club, then for the club who both put on a mark-up to the shooter and sold those killed to a dealer, who finally sold it on to become a pigeon pie. The small birds too would have, without question, ended up under a crust, because it was common practice to net roosting sparrows out of the thatch to make sparrow pie. Before we leave the culinary side of live pigeon trap shooting, reference needs to be made to what was termed 'by-shooting'. Some clubs did not permit the practice but, at others, shooters gathered outside the club grounds and attempted to bag those birds not killed in the ring. With a dead bird valued at 3d (1½p) or 4d (2p) and a cartridge costing but a 1d (½p), there was sport

---

The following are a few of Mr. W. T. Coton's Pigeon Shooting Matches:

The first win was at the age of twenty-one, when at Stratford-on-Avon in 1883 he won a bullock. After that, a list as follows: Divided a bullock at Northampton with Roger Tichborne, then divided twelve bullocks with E. Lovell, also at Northampton. Following that, shot a match for £50 with Roger Tichborne, at Witton, Birmingham, and won. Then divided a bullock at Northampton, won a bullock outright at Nuneaton, divided a bullock in Coventry Park, then another at Coventry, one at Radford, one at Bletchley, one at Four Ashes, then another at Radford. Won a bullock outright, then a match at Whitley with Capt. Fowler for £50, and beat him. Won a gun at Birmingham, another at Coventry. Won a fat pig at Nuneaton, also divided a pig at Four Ashes and another at Bletchley. Then got first, second and third in a £52 Starling handicap at Witton, Birmingham. Following that, first and second in another Starling handicap at Witton, this time winning £48. Then in August 1896, at Brighton, won the Grand International Cup, valued £75 with £25 in money added, from scratch, beating a field of one hundred and three competitors, killing seventeen birds out of eighteen, and defeating some of the world's best shots; the next year running up second in the same event, killing fifteen birds out of seventeen. At the Union Gun Club at Hendon, won a second prize (£117), killing fourteen birds out of fifteen; also another second prize of £117 at Hendon, killing thirteen birds out of fourteen, off 29 yds. mark; second again at Hendon, winning £115.

Mr. Coton is also the winner of upwards of twenty cups, and has shot seventeen matches, winning fourteen and losing three.

The whole of the above have been won by cartridges loaded by Mr. W. T. Coton himself.

Advertisement issued by W. T. Coton, specialist cartridge loader of Coventry

---

to be had and yet another profit to be made.

On a rather different scale were the profits generated for the gunmaker. The timing of the upsurge in fashionable interest in the pastime meant that this was another class of shotgun that evolved over the second half of the nineteenth century. Interestingly, the changes in the pigeon gun did not exactly mirror the game shooting gun, reflecting the different roles of the two types. Almost from the beginning of live pigeon trap shooting, special guns had been produced for the devotees of the 'sport' and this trend continued into the second half of the century. Until the mid-1870s, a fair but diminishing number of muzzle-loading guns continued to be used, reflecting the popular view that a well-loaded muzzle-loader was

Charles Boswell, London. Twelve-bore live pigeon gun. Anson & Deely action

a more efficient shooting gun than the new breech-loader. In the early days of breech-loading, some pinfires were used, but they did not last as long at the traps as their equivalents did in the field. By the early 1870s, the guns were either centre fires or muzzle-loaders, by the early '80s, the choke-bored centre fire reigned alone.

One feature of the Hurlingham rules had been to outlaw the 'young cannons' and unrestricted shot loads that had been a feature of pigeon shooting at the beginning of the century. The effect of the new restrictions was to channel the development of the pigeon gun towards the most efficient use of 1¼oz of shot in a 12-bore barrel. The choke bore, as has already been mentioned, was enthusiastically adopted while, at the other end of the barrel, another set of changes took place. While the shot charge was regulated, at first the powder charge was a matter of choice and inevitably the value of a high-velocity load was soon realised, but by the 1880s the powder load was regulated also to four drams. So, to accommodate a large charge of powder, 1¼oz of shot and an efficient wad column, the need was for a longer cartridge case. First to appear was the 2¾-inch case, instead of the game shooters' 2½-inch, followed by the 3-inch and ultimately the 3¼-inch.

To handle these potent cartridges, and most of all to tame their recoil to give the shooter a chance to make efficient use of his second barrel, the need was for a heavy gun. Under the Hurlingham rules, there was an upper limit of 7½lb, but other venues had either different rules or none at all. This, of course, was good news for the trade, because a keen shooter might well be induced to have one gun for the Hurlingham rules and another for less controlled events.

The same need to withstand the repeated use of the heavily-loaded, high-velocity ammunition naturally resulted in gunmakers paying special attention to the strength of the actions of their pigeon guns. Some of the features such as 'side clips', where the standing breech has two small forward-projecting lips between which the barrels fit, were probably of more value in the mind of the user. However, some sort of extension to the barrel rib which became the bearing surface for a cross bolt or another third bite was, if well fitted, a real contribution to the strength of the gun. Given the more deliberate mode of shooting at the traps, these top extensions were less of a nuisance to the user than the same additions to a game gun.

Another difference between a game gun and a pigeon gun was often the rib between the two barrels. There were wide differences of opinion on the subject of barrel ribs, but speaking generally, as something more like deliberate aim was taken with a trap gun, the broad, file-cut, non-reflecting rib tended to be more usual.

A more subtle difference between the two varieties of shotgun lay in the stocks. Since the mark was almost invariably rising, the stock on a trap gun would be straighter and shorter than that on a game gun, as these two differences would tend to make the gun throw the centre of its pattern above the point of aim.

The sale of these specialist guns was, as far as the gunmaker was concerned, only part of the story. Of great value to him was the potential for the pastime to promote his wares. Because of the prominence of the participants their doings were widely reported and this, of course, would include the name of the maker of the winning gun. As a result, in the days of muzzle-loaders it became common for certain gunmakers to be present at important matches to load their clients' guns, and, since there was not a word in the rules to prevent it, to encourage and coach their clients at the same time. In this way the reputation of Messrs Purdey was maintained, while it provided an *entrée* to more recently established gunmakers, such as Grant.

Another way in which a gunmaker could gain kudos from live pigeon shooting was to win the prize himself. Inevitably, this was not an option open to all, for a pre-condition was that the gunmaker should also be an excellent shot, but it was the route followed by Edwin Churchill and Charles Boswell. Both were men of relatively humble backgrounds who, by the turn of the century, had gun-making businesses in central London, an example followed to a lesser degree by some provincial makers.

As a result of these reputations, pigeon guns tended to become the preserve of certain makers. Not that there was any reason why other

Carr Bros., Huddersfield. Advertisement from *The Shooting Times*, 1898

Charles Boswell hammer live pigeon trap gun

gunmakers could not have built a gun of the same specification, as indeed they did, but the specialists had a firm grip on the business. To consolidate their position further, these makers recruited the best shots. There was nothing particularly underhand about this, indeed Boswells advertised 'very special terms to good shots for new guns'. The actual price paid, if there was one at all, was, of course, a matter for negotiation.

The advent of smokeless powder in the 1880s and more especially in the 1890s brought into pigeon shooting another group of traders who could derive advertising copy from the events. Soon the total prizes and events won became

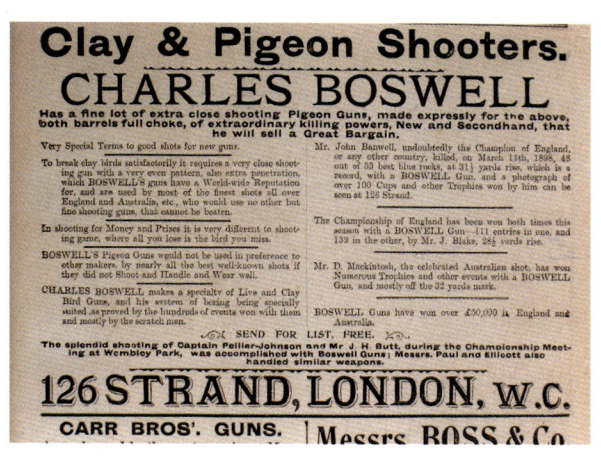

Charles Boswell, London. Advertisement from *The Shooting Times*, 1898

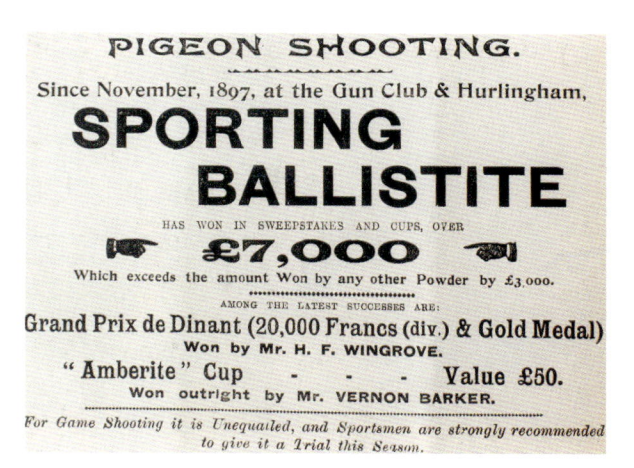

Advertisement for Nobel's Explosives Co. Ltd, from *The Shooting Times*, 1898

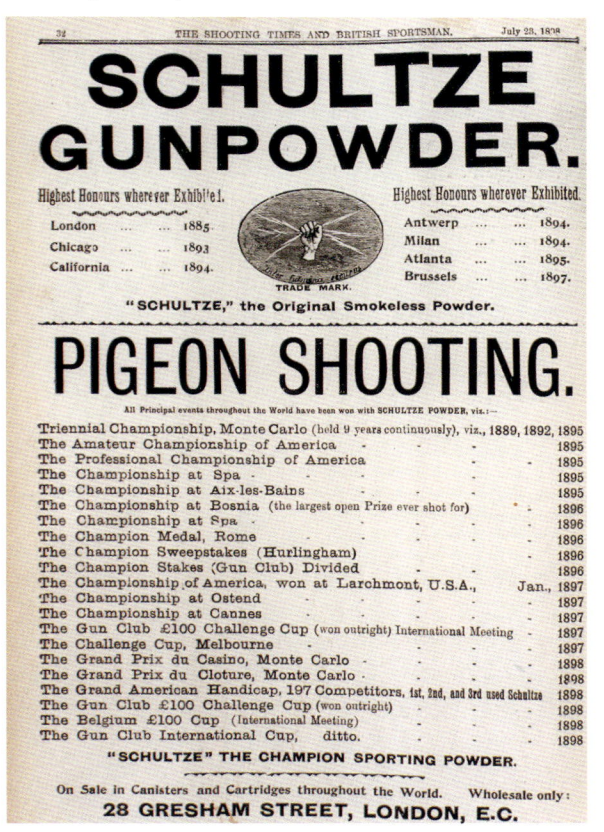

Advertisement for The Schultze Gunpowder Company from *The Shooting Times*, 1898

important features of their advertising and, like the gunmakers, they tried to corral crack performers. Doubling the prize money was one way of attracting the talented, but there were obviously many other lures cast, most of them unreported.

Such practices, and the gambling that was also a feature of the scene, gave live pigeon shooting the reputation of being a decadent money game and this tainted image served to hasten the decline of the 'sport'. Indeed, from the 1870s onwards, there were serious attempts made to outlaw the practice. In any case, support was dwindling. For instance, on the opening day of the Gun Club season in 1901, only six gunners turned up to shoot, while at the prestigious Hurlingham Club a couple of days later only four competitors shot.

By contrast, the season at Monte Carlo, which lasted from December to April, was well attended with a clientele drawn from across Europe, no doubt attracted by tempting prizes. The top English prize winner in the 1900–1 season was the Hon. R. Beresford, with a total of £518.

Right from the very beginning there had been opposition to live bird trap shooting. Back in the beginning of the nineteenth century, Col Hawker had roundly asserted that 'pigeon shooters could spare innocent blood by shooting at shied pennies'. The problem was that insufficient effort had been given to discovering any credible alternative to a living bird. Moreover, when inanimate targets did come to be devised, the motivations were as much convenience and cost as ethics.

Somewhat surprisingly, the first shotgun target

to enjoy any wide usage was the glass ball, originally produced in Great Britain, but vigorously promoted, and indeed patented, by the American professional shooter, Adam H. Bogardus, 'Champion Wing Shot of America'. Incidentally, the basis for his patent on a glass sphere was to protect the idea of ridges moulded in its surface which, it was claimed stopped the shot glancing off it. The Bogardus ball was about 2¾-inches in diameter, roughly the size of a cricket ball, and it was thrown from a trap consisting of a cup on the end of a flat spring, which, when released by a simple catch, flipped the ball up into the air.

While it is true that a glass ball smashed in a very satisfying manner and a few chicken feathers stuffed inside it would float down and create the illusion of a bird shot in the air, the litter of glass fragments this produced was a dangerous nuisance. Naturally, all sorts of alternatives were tried, such as balloons in cages and metal spheres filled with powder which puffed out when hit. For the exhibition shooting part of Col Cody's Wild West Show, which came to London at the time of Queen Victoria's Golden Jubilee in 1887, H.A.A. Thorn produced a mould to cast a hollow pitch ball which could be launched by a Bogardus trap, thus safeguarding the bare feet of Red Indians and the hooves of horses as they dashed about the arena.

A glass ball lobbed up from a spring trap is not a very testing target and it was found possible to shoot huge numbers in exhibitions. Inevitably, the need was for something more testing and this was found in what could be described as a species of mechanically-thrown discus made of some frangible material. This time the true inventor was an American, George Ligowsky of Cincinnati, Ohio, who obtained a patent on the idea in 1880. His first targets were baked clay, almost exactly the same colour as terracotta which, unfortunately, were often almost impossible to smash with a fair hit. There followed a search for a cheap material combining the strength necessary to stand the shocks of travel and projection, but which would shatter when hit. All sorts of compositions were tried and the first successful one was a mixture of river silt and plaster of Paris.

In view of the problems besetting the sport of live pigeon trap shooting, it is not surprising that there should have been thoughts on using the Ligowsky bird for the purposes of competition, and by 1884 the first clay shooting clubs had been formed in Great Britain. A more cynical view is that clay bird shooting was promoted by gunmakers as a means of selling more cartridges. At live pigeons the rule was commonly, one miss and out, whereas at clays, the whole round would be shot. Moreover, if a new shooting public could be created, gunmakers would sell more guns as well as more cartridges. A more balanced view holds that, in essence, public opinion and gunmakers' efforts were acting in the same direction and, therefore, they naturally combined to produce the phenomenon of competitive inanimate bird shooting.

Advertisement for Cogswell & Harrison Swiftsure Trap from *The Shooting Times*, 1898

SHOOTING TOURNAMENT!

The Clay Bird Shooting Association's 14th Annual Open Championship Meeting will take place at

THE MIDDLESEX GUN CLUB, HENDON, N.W.,

On Thursday, Friday, and Saturday, June 21st, 22nd, and 23rd, 1906, from 11 a.m. to 7 p.m. daily.

PRIZES VALUE £564 10s.

This Annual Meeting affords a splendid opportunity for sportsmen to compete or to witness shooting at clay bird- by competitors from all parts.

THURSDAY'S PROGRAMME.

Time.	Event.	Entrance Fee.	Value of Prize.	Handicap or Scratch.
11.0 a.m.	Ladies' Prizes (4)	7s. 6d.	£15 10 0	Handicap.
12.0 noon	R.S.P.C.A., 4 prizes	5s. 0d.	£12 5 0	do.
2.0 p.m.	"County Gentleman" (No. 1), 7 prizes	5s. 0d.	£66 10 0	Scratch.
3.0 p.m.	Dewar Inter Club, 7 prizes	8s. 0d.	£20 10 0	do.
3.30 p.m.	Brussels Gun Club 1 prize	None	£15 0 0	do.
3.30 p.m.	Westley Richards, 4 prizes	6s. 0d.	£18 7 0	Handicap.
4.30 p.m.	"Shooting Times," 5 prizes	5s. 0d.	£21 0 0	Scratch.
5.30 p.m.	Cooppal, 4 prizes	4s. 0d.	£13 10 0	Handicap.

FRIDAY'S PROGRAMME.

Time.	Event.	Entrance Fee.	Value of Prize.	Handicap or Scratch.
11.0 a.m.	Musgrove, 4 prizes	5s. 0d.	£13 0 0	Handicap.
12.0 noon	Vice-presidents', 4 prizes	7s. 6d.	£22 10 0	do.
2.0 p.m.	"Field" 4 prizes	7s. 6d.	£22 10 0	do.
3.0 p.m.	Walkers, Parker, 4 prizes	5s. 0d.	£16 12 0	do.
4.0 p.m.	Middlesex Gun Club, 5 prizes	7s. 6d.	£28 6 0	do.
4.45 p.m.	£20 Specie Event, 6 prizes	15s. 0d.	£20 0 0	Scratch.
5.30 p.m.	"County Gentleman" (No. 2). 5 prizes	7s. 6d.	£23 5 0	Handicap.

SATURDAY'S PROGRAMME.

Time.	Event.	Entrance Fee.	Value of Prize.	Handicap or Scratch.
11.0 a.m.	Marylebone, 4 prizes	5s. 0d.	£15 10 0	Handicap.
12.0 noon	International Shield, 9 prizes	12s. 6d.	£95 0 0	Scratch.
2.0 p.m.	£50 Specie, 7 prizes	15s. 0d.	£50 0 0	Handicap.
3.0 p.m.	Championship 6 prizes	12s. 6d.	£64 15 0	Scratch.
Final	Grand Aggregate, 2 prizes	5s. 0d.	£10 10 0	do.

Shooters who are not members of affiliated clubs will be charged 1/- extra for each event. The prizes will be distributed on Saturday afternoon, June 23rd, at 5.30 p.m., by Sir John R. Heron-Maxwell, Bart.

Band Last Day.   Refreshments on Grounds.   No Betting.

Livery and Bait Stables and Bicycle Storage at Hotel opposite grounds. Admission 2/6 each day (returned to shooters).

Programme post free on application to A. H. GALE, hon. sec., 178 New Bond Street, London, W.

We have received an advanced copy of the programme of the Clay Bird Shooting Championship Meeting, which

Advertisement for the 14th Annual Open Championship of The Clay Bird Shooting Association from *The Shooting Times*, 16 June 1906

Inevitably, there was much discussion of the new pastime. Critics made much of the fact that, while a trapped pigeon or wild bird usually presented a rising and accelerating mark, by contrast inanimate targets were always slowing down and very likely falling as well. So, the discussion turned on the advisability of stocking the gun to take account of this characteristic, bearing in mind the fact that the typical live bird gun was set to shoot high.

A new twist was given to the question of the best gun to use by the outcome of the 'Great Anglo American Match' held at the Gun Club at Hendon in June 1901 with prize money totalling £1000.

It has to be said that the rules of this contest were curiously drafted, presumably to take account of the different style of clay pigeon shooting practised across the Atlantic. The US shooters were only permitted a single shot at each clay, but were allowed to use 1¼oz of shot. By contrast, the home team could use both barrels, but was limited to the usual British clay shooting charge of 1⅛oz. The snag was that the rules made no stipulation as to how the gun was to be held prior to calling for the bird. So, while the English shooters held their gun stocks below their elbows, their rivals bedded their stocks firmly to the shoulder and took something like an aim before calling.

Overall, there were clear signs that the newly-formed Inanimate Bird Shooting Association was making determined efforts to strike out in a new direction and distance itself from the ways of the live bird trap shooters. Those who were attracted to the new sport, were largely drawn from the growing bicycling class. £15 to £20 for a new bicycle cost almost as much as a middle grade gun, nevertheless contemporary photographs show racks of cycles behind the firing line. An outing to the traps was another expression of the new freedom that came with personal transport. The most immediately noticeable change was the banning of gambling which had always been a feature of live bird matches. There were other attempts to attract a different clientele, for instance, the entry fees for the competitions were lowered to 7/6d (37½p) for national competitions, compared to £2 or £3 , which were the normal sweepstake entry fees at Hurlingham or The Gun Club. However, 7/6d (37½p), plus an extra 1/- (5p) if you were not a member of an affiliated club, must have been a deterrent to a casual entry by a farm labourer for whom this represented over half a week's wages. The stipulation of a maximum load of 1⅛oz of shot was also important in this, because such cartridges were significantly cheaper than the specialist 1¼oz loads of the Hurlingham rules.

The account of the international pigeon match describes the winning American guns as 'heavy

Four photographs from *Land & Water*

live pigeon shooting guns' and so highlights something of the continuity in gun design. Notice also the Boswell advertisement that stresses the need for guns for both forms of shooting to throw tight but even patterns.

By contrast with this effortless carry-over of expertise by the gunmaker, the gunners did not seem able to make the transition in the same way. It is a vain search to try to find many live bird shooters in the ranks of the winners at clays. The answer to this conundrum seems to be that, while to the layman these two forms of shooting may seem very similar, they were, in fact, very different. One noted performer at live birds claimed that he found clays too easy. However, the account of even the £1000 match in *Land and Water* starts with the statement that 'a claybird contest is not very exciting'. It would seem that two different temperaments were needed, something of mercurial brilliance to shoot a dozen unpredictable pigeons, or a more calculating concentration to show well at a hundred more routine clays.

THE I.B.S.A. CHAMPIONSHIP TOURNAMENT.
Shooting for THE COUNTY GENTLEMAN Cup.

CHAPTER NINE
A Gun for a Wildfowler

There was a time when the annual southern migration of huge flocks of wildfowl was regarded as an event that proved the existence of divine providence. Coming as it did at the leanest time of the year, this wholesome addition to a meagre diet was greatly appreciated.

So, from time immemorial, ducks and geese had been netted, trapped and snared to be sold at good prices in the markets. When the gun made its appearance, it was just another means of taking fowl. Indeed, as long as commercial fowling existed, it never completely supplanted the net.

As far as can be ascertained, the crude guns of the professional wildfowler were the first shotguns and must, therefore, be regarded as the original from which all other variants have evolved. This fact is enshrined in the generic term of 'fowling piece' for all shotguns.

Centuries of hunting had caused selective breeding of wary fowl and so, right from the first shot, long range was required. That is not to say

A bag of widgeon shot with a hammer gun by J. Purdey

Flight shooting from behind a screen. Sidney Smith. (PHOTOGRAPH COURTESY OF THE YORKSHIRE MUSEUM)

that, under favourable conditions, closer shots were not to be had but, on balance, out on the marshes and foreshores, the long shot was the norm.

In view of this fact and to explain why the wildfowler's gun evolved as it did, it is necessary to give a brief outline of shotgun ballistics. This fascinating branch of physics is complicated by a host of variables which, at times, compensate for each other but, at others, are cumulative in their effects. For the sake of simplicity, we can say that for a gun of a given size, the only variable really under the control of the gunner is the size of his shot. With the smallest sizes, the cloud of shot flying through the air is dense but, because the shot is small, its energy is soon spent. With larger shot, the converse is the case. So the conundrum facing the wildfowler is that although with large shot he has the potential to kill at very long ranges, because a given weight of charge will contain fewer pellets, the gaps in the shot cloud at long range can easily accommodate a duck.

To a degree this difficulty can be overcome by firing at a flock of ducks, but then we introduce problems even more intractable than those of ballistics; those of ethics and economics. Ethically, in flock shooting the scattered shot cloud is just as likely to wound as it is to kill cleanly and economically, a professional fowler needs to see a return for his powder and shot.

Fortunately for the purposes of this consideration, by 1900, flock shooting was regarded as unsporting and advertisements and readers' letters extolling it were being derided in the sporting press.

If we define as a sporting load a charge which can be expected to kill cleanly the single bird aimed at, we still have the need to balance the energy of each pellet with the density of the shot cloud. Since long range is required, the best option lies with a large charge of shot, in which each pellet is only just large enough to retain sufficient energy for the distance, flying close enough to its neighbours to ensure that several pellets strike.

Long experience dictates that the shotgun needs to weigh about 6lb for every ounce of shot in the charge. So the next variable that has to be accommodated is the maximum weight of gun the fowler wants to carry. Even for a strapping young

J. & W. Tolley 4-bore single-barrel, with leaf sights for long range shooting

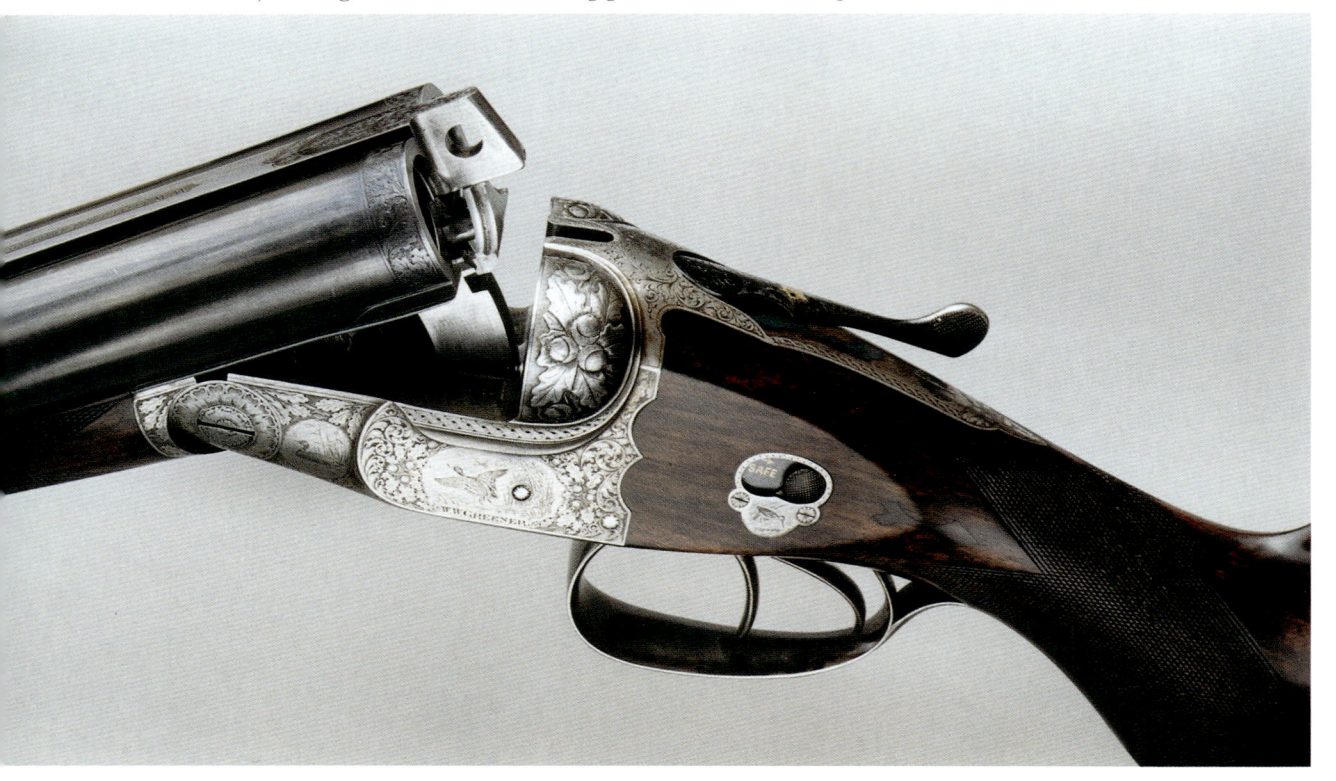

W. W. Greener, Birmingham. 8-bore wildfowling gun with Facile Princeps action

man, 20lb of gun is about the limit. This translates to a shot charge of just over 3oz, which is the usual load for a 4-bore cartridge.

For those of more mature years or lesser stature, a lighter gun is the only option. Long experience has taught that a happy compromise is a gun of some 12lb firing about 2oz of shot, and this translates as the specification for an 8-bore shotgun. When such a gun is well choked, it is perhaps the best option for the foreshore. Next down the scale comes the 10-bore, weighing in at some 9lb, firing a proportionately lower charge, but still a potent gun and the favourite of many wildfowlers.

Since relatively few shots are fired on a fowling expedition, there was no real need on a big gun for many of the improvements which were such an integral part of the nineteenth century shotgun. In general, therefore, the fowler's gun remained a simple, screw-grip underlever hammer gun with

W. W. Greener, Birmingham. 8-bore detail of trigger guard

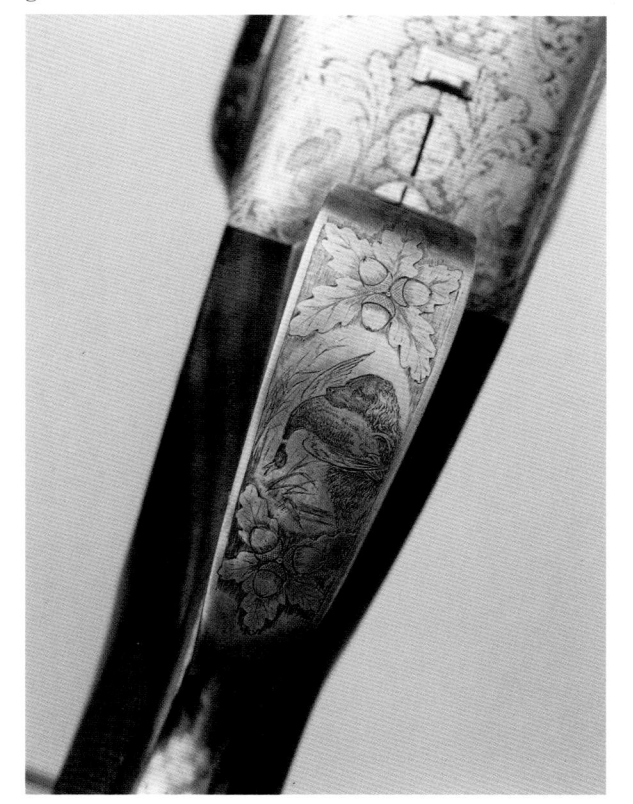

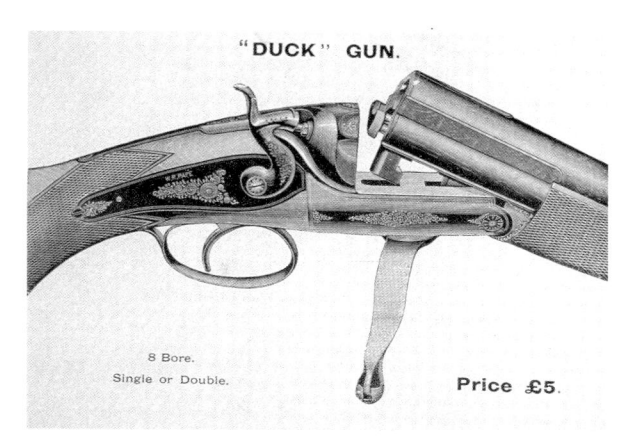

W. R. Pape, Newcastle. Page from catalogue, mid-1890s

the choke bore the only acknowledgement of the feverish development surrounding the game gun. This is not to lose sight of the fact that 10-, 8- and even 4-bore guns were made for gentlemen fowlers with all the latest improvements, but they were much more a gunmaker's *tour de force* than any response to the practical needs of a long-shore gunner.

All gunmakers, because they could order from the trade in Birmingham, could supply large calibre wildfowling guns and there was usually a brief notice in the retailer's catalogue to this effect. However, some firms made more of an effort to capture the fowler's trade and, in this respect, the firms of Thomas Bland in London along with J. & W. Tolley and W.W. Greener Birmingham were probably the best known. It should be noted that while Blands' headquarters and retail premises were in King William Street, Strand with works in Whittall Street, Birmingham, Tolleys had a showroom in New Bond Street, London, but their base was at the Pioneer Works in Loveday Street, Birmingham. Likewise Greeners were also Birmingham-based but with London showrooms.

Of the three, Tolleys were probably best regarded as specialist wildfowling gunmakers. There is the delightful tale of a man telling an old woman in the west of Ireland that he proposed to go duck shooting. The crone promptly advised him to take a 'Tolley-gun', said in such a way that it was evident she had no idea what the term meant.

The typical Tolley gun had back-action locks and was decorated with nothing more than a very restrained border engraving. Some of the actions were nickel-plated as a further protection against the salt, the inescapable hazard of the foreshore, but beyond this there was nothing which could be described as embellishment. Yet the typical Tolley is an attractive gun, the result of skilful action filing and stocking, and distinctive because the size of even a 10-bore is such that it cannot be merely a scaled-up game gun.

The demand for large calibre guns was inevitably small compared with smaller game guns. Also, given the virtually static design, there was less inducement for gentlemen fowlers to replace their guns. This and the fact that, because of their size, they were special in all their parts, meant that the manufacture of such guns would have been the province of relatively few hands in Birmingham. So it is most likely that some of the guns found with other names are, in fact, Blands, Tolleys or Greeners sold to the trade.

One way in which the balance of the equation of weight versus its user's ability to wield a mighty gun could be tilted in the fowler's favour was to employ some sort of rest for the gun. This compromises the ability to swing the gun rapidly in any direction, but, in return, the shooter can use a piece of far greater weight propelling a far greater charge.

This is a notion of great antiquity. In war the musketeer of the sixteenth century rested his ponderous weapon on a forked stick, while the contemporary fowler used what was accurately called a 'bank gun' to shoot into a flock of feeding or resting waterfowl. From there it is but a short step to mount the same or an even larger gun on a shallow-draft boat and so be able to move the great gun more easily and reach birds away from the margins of the land. The result, in embryo at least, is what we have come to call a 'punt gun' mounted on a gunning punt.

It is probable that there is no sporting gun more misunderstood than the punt gun. When a shooter, even one with great experience with a

T. Bland & Sons, London & Birmingham. Advertisement from *Land & Water*, 1897. Notice sizes of pattern quoted

---

BLANDS' GUNS

REPORT OF
SIR RALPH PAYNE GALLWEY'S
TRIAL OF
WILDFOWL GUNS.
See "FIELD," Nov. 27, 1897.

"SINGLE BARREL, 4-Bore.
Charge, 10 drams Powder, and 3¼oz. B Shot.
Average Pattern, 60 Yards.
Target 12ft. by 12ft., 264.
Within a Circle 4ft. Diameter, 135.
A fine pattern, and a most suitable one for groups of Wildfowl."

"DOUBLE BARREL, 8-Bore.
Charge, 6¼ drams Powder, 2¼oz. B Shot.
Average Pattern, 50 Yards.
Target 12ft. by 12ft., 187.
Within a Circle 4ft. Diameter, 130.
This Gun, like the 4-bore, made excellent patterns and just the ones for small groups of Wildfowl."

"DOUBLE BARREL, 10-Bore.
Charge, 4 drams Powder, 1½oz. B Shot.
Average Pattern, 50 Yards.
Target 12ft. by 12ft., 115.
Within a Circle, 4ft. Diameter, 81.
A Gun of this description is well adapted for bringing down large single birds, as wild geese, or for firing at a few fowl grouped together."

SIR RALPH PAYNE GALLWEY CONCLUDES HIS REPORT BY SAYING:
"*It will be seen that Messrs. Bland and Sons' Guns shot most admirably, particularly the 4-Bore and the 8-Bore.*"

WILDFOWL GUNS.
BEST MATERIALS AND WORKMANSHIP.
SHOOTING GUARANTEED AT EXTREME RANGES.
FITTED WITH RECOIL PADS.
4-Bores, Single, £25.
8-Bores, Double, £25; Single, £16.
10-Bores, Double, £15; Single, £10.

T. BLAND & SONS.

A GUN FOR A WILDFOWLER

Lock from typically crude East Anglian fen muzzle-loading punt gun

Dr William Kingdom-Murrill sculling to fowl in an area known as 'The Kench' in Langston Harbour, Portsmouth.
(PHOTOGRAPH COURTESY OF N. HORTON)

HEYDAY OF THE SHOTGUN

Punt gun 1½-inch cartridge 8 inches long, compared with a 2½-inch 12-bore

Charles Boswell London. Double barrel 8-bore 4½-inch chambers 48" barrels weight 21 lbs. If both locks are cocked the rear trigger discharges both barrels

game gun, is told of a species of shotgun with a barrel 8½ feet long with a bore of 1¾ inches from which ¼lb of black powder propels 1¾lb of coarse shot, he will instantly think that the piece has enormous ranging power and is totally unsporting. The truth is very different. The laws of ballistics, principally the effect of atmospheric resistance on the spherical lead shot, limit even the heaviest loads to about 90 yards. As with a shoulder gun, it is possible to load very large sized shot and make fluke kills at huge ranges.

To manoeuvre the gun to within 90 yards of a flock of wary ducks on open water needs great skill. Moreover, since the most effective shot is one made just as birds rise, which alert members of the pack do as they see the smoke from the gun, it follows

Breech of 1¼-inch gun converted from a muzzle-loader, wire-wound at the breech for extra strength at the time of conversion. (PHOTOGRAPH COURTESY OF N. HORTON)

that the gun's muzzle has to be elevated as it is fired. Certainly gunners of great skill and experience have, in very favourable conditions, made huge bags but the average is surprisingly modest – especially when the effort, not to mention the danger, inherent in each trip is considered.

The technical revolution had even less to offer the punt gun than the shoulder wildfowling gun. So the punt gun remained one of the last strongholds of muzzle-loading, but even here breech-loading became accepted eventually. The reasons were subtly different from those that had influenced the game shooter. Speed of use was certainly not needed in a punt, where a couple of shots per trip would be the norm. To the punt-gunner, a breech-loader offered the chance to change loads if different quarry or conditions presented themselves or to unload at the end of a trip.

When breech-loading did come, the mechanisms used owed much more to the design of a piece of artillery than to a sporting gun. Not only the sheer size of the gun, but the lack of room in a shallow punt, made a screw-in breech a practical and, with well-made buttress threads, a very strong option.

However, there were various forms of drop-down action, from a compact block not much longer than it was wide, typically on Holland & Holland guns, to guns that looked like large calibre shoulder guns with their stocks chopped off.

Developing this theme a stage further, there were guns that bridged the punt/shoulder gun gap by being adapted to either role. Essentially, these were the biggest of shoulder guns with a hole for a

"NORMAL" IN "NIMROD"

WONDERFUL POWDER!
WONDERFUL CASE!

Erith Flood. — "NIMRODS" loaded with **NORMAL POWDER 15 days** in water—result:

Powder perfectly **DRY**, without the slightest deterioration.

Cases only slightly swelled—entered perfectly into Gun.

Caps perfect, and no misfires.

Pressures and Velocities equal dry Cartridges.

BEAT THIS!!

The Normal Powder & Ammunition Co., Ltd
38 & 39, PARLIAMENT STREET, LONDON, S.W.

Opportunist advertising from *Land & Water*, 1897

From *Shooting Times* 1898

Kynoch tool to crimp thin brass cases

breeching rope bored in their stocks sitting in a specially-made cradle mounted on the punt.

Damp is a virtually inseparable part of wildfowling and, while all the fowler's gear is affected and chosen to minimise the effects, breech-loading cartridges with paper cases were among the most vulnerable. There is nothing quite so frustrating as a gun that cannot be loaded or unloaded out on the marsh with birds coming over. In fact, by 1900, the best paper cases were remarkably water resistant, witness the opportunist advertising of the Normal Powder Company after their premises had been flooded.

However, some twenty years earlier, there had appeared on the British market cartridge cases that were either brass-covered paper or made entirely of brass. These developments were a natural application of the techniques perfected in the 1870s for manipulating brass into cartridge cases for military small arms, especially machine-guns. The brass-covered paper case still had a vulnerable paper roll-turnover, but was especially suited to the smooth functioning of the new ejector guns. By contrast, the all-brass case, called 'The Perfect' by its makers, Messrs Kynoch, was closed by having the open end folded into a cone-shaped crimp. Finished with a touch of wax at the

John Rigby, London. 10-bore, thin brass case, wildfowling gun, built on Rigby & Bissell's patent action (1879)

apex, it was totally waterproof and so of obvious interest to the wildfowler.

Like so many technical advances, the brass case was not pure gain and had its disadvantages. Its thin walls necessitated a larger-sized wad, roughly speaking a gauge size larger than the nominal size of the case. This, in turn, needed a barrel bored to the new size, so producing a specialist gun that would not fire efficiently the paper-cased ammunition that fitted its chambers.

However, the thin cartridge walls had two advantages to offer. The more obvious one was that the cartridge case had a greater capacity than the equivalent paper case. The less obvious one was that because there was less difference in size between the outside diameter of the cartridge case and the bore of the gun, the cone that was the usual way of connecting chamber to bore could be almost eliminated.

The importance of this seemingly minor change lies in the fact that the deformation inflicted on a charge of lead shot as it is squeezed under high pressure from the cartridge to the bore, is reduced.

J. & W. Tolley, Birmingham. Rib inscription on 'Altro' gun

These features of the brass-cased cartridge combined to enable the production of a most effective gun. What it appeared to be was a 12-bore with the power of a 10-, or a 10- with the power of an 8-bore. The brass cases were not cheap. For instance, the 1902/3 Kynoch catalogue lists 2½-inch 12-bore Perfects at £3/17/6d (£3·87½p) per thousand, against £2/10/0d (£2·50p) for Kynoid waterproof cases, or £1/12/6d (£1.62½p) for the cheapest 'Brown' quality case. This premium price was offset by the much longer reloading life of the brass cases, provided they were looked after and washed out with a weak washing soda solution soon after use.

The thin brass case had been available from the 1880s but, in 1903, the first results of some work that gave new potential to this type of ammunition were published. The announcement was in a long letter to *The Field* of 28 February:

GUNS WITHOUT CARTRIDGE CHAMBERS.

Sir, – We have nearly reached the end of the shooting season, and it may interest some of your readers who go wildfowling, and have experience of the fatigue and annoyance of carrying and using the larger wildfowl guns, to know of the power and handiness of brass-case 12-bores (i.e., 10-bores) with and without cartridge chambers, weighing under 8lb, and with 30in barrels. I take it most of us are aware that the conditions prevailing about the chamber of a 12-bore gun when firing a paper cartridge are not ideal for hard shooting, but they are convenient for the purpose of game shooting and where rapid loading is necessary, and somehow the same system has come to be used for all bores and purposes; but for wildfowling, where the greatest power is desirable, it is not the best, and I do not think the paper case plan is the one best adapted to lessen recoil.

I take it that, when a paper (or brass-covered paper) cartridge is fired, some of the power of the explosion is used up in stretching the case until it fits the chamber tightly, some of the power is expended in compression of the paper itself, some is wasted by the friction of the wads and shot against the rough paper of the cartridge, and enormous must be the friction of the shot against the cone, for, by the time the shot leaves the cartridge, the stretching and compression of the case has probably enlarged it to 11-bore, and the shot, on leaving this 11-bore cavity, can slightly expand laterally before it touches the cone, and then in the next ½in (for the proximal part of the cone is probably very little used, and the cone is only ½in long) it is compressed into a diameter of about 13-bore (i.e., the usual 12-bore gauge). Besides the loss of energy in overcoming the frictions I have just named, there is damage to outside shot against the cone, so that, on leaving the muzzle, these damaged shot may fly off at a tangent and be dangerous; anyhow, they are lost to the killing circle. To overcome the causes of friction I have indicated must use up a considerable portion of the propelling force of the powder gases, and I consider this probably explains the need that paper-case guns have of a larger charge of powder in proportion to shot than either muzzle-loaders or brass-case guns, which have practically no chamber.

Last summer I had a single-barrel experimental gun made to take the thin 12-bore brass case. I used the French one, as it is solid drawn, practically a true cylinder, of less brittle brass, and fitted with a better cap for nitro-powder than any English case. The Germans have found out the good qualities of this cap, for they are fitting it, though of French make, to cartridges of their own make for use at Monte Carlo and elsewhere. I have used this cap for ten years – that is, ever since they and the French cases were exhibited in London. This single-barrel gun I had built without any chamber, and to ensure hard shooting I used 9-bore wads over the powder, and it did so well at the target that I ordered a best double to be built on the same plan, and this double gun, though only 7½lb weight, will shoot 2oz shot comfortably and with good penetration. I get a pattern of 170 No. 2 shot on a 30in circle at 40 yards very well centred; but the important part of the loading is the small amount of powder that is needed, for 1½oz shot needs only the equivalent of 3¼drms black powder, and 2oz. shot needs barely 3½drms (38 grains E.C. No. 3 was used for the 2oz charge).

I believe our ancestors shot game with 2oz of small shot in small-bore muzzle-loading guns of 8lb weight, of course with black powder, and with barrels neither too smooth nor clean inside. Surely we ought to be able to do so with nitro powder, clean barrels, and bigger bore, and it was the confidence I had of being able to do so that made me spend the time on the experiment. I found, also, later that a gun I bought second-hand just before Christmas, and built much on the same lines, except that there was the faintest possible chamber, shot very hard and close using 10-bore wads, but it has not the advantage of the chamberless gun of taking long or short cases equally well.

At Christmas in Ireland I used this gun with thin chambers (7lb 14oz), as the double was not ready, and I did the best long-range shooting at duck, coot, &c., I have ever done. Of course, it is impossible to say the range always, but from the enormous

HEYDAY OF THE SHOTGUN

J. & W. TOLLEY'S
LONG RANGE
WILDFOWL GUNS
Single and Double 10, 8, 4, and 2-Bores.

The Editor of the "FIELD" says:
"Messrs. J. and W. Tolley, the well-known firm, have for some years paid great attention to wildfowl guns. . . . Their 8-bores flatten a BB shot to the size and thickness of a sixpence at 100 yards or more."

From Sir R. PAYNE-GALLWEY, Bart., Thirkleby Park, Yorks:
"I used your 4-bore for the first time the other day in Ireland, and got in one shot seven widgeons, four teal, and two curlew. I consider it invaluable."

From H. R. SHIELD, Esq., Parkglas, Narberth:
"Nothing could be better fore shore shooting than the 8-bore Gun you sent me. I can now believe in a gun that is safe of a single bird at 90 yards, and a flock at 130 or over. I killed shore birds at 150 yards the other day."

From Hon. E. WILLOUGHBY, Settrington House, York:
"I like the 8-bore very much. I killed a partridge dead yesterday at 100 yards, and also some plovers—very long shots."

From J. P. GARNETT, Esq., Willington Manor, Bedford:
"I have given the 10-bore Double a good trial, both at ducks and plover, and like it immensely. I have shot plover at 120 yards with it, firing into a flock."

From the "FIELD," October 23, 1897:
"I am using one of Messrs. J. and W. Tolley's 'Altro' guns now for the second season, and its power is simply astonishing. I have not found that changing from an 8-bore to a 12 has at all diminished my sport among the ducks, for your gun shoots so admirably that one is able to command almost the same range. For snipe shooting the gun is just as handy as a light paper case 12-bore."—FLEUR-DE-LYS.

From the "FIELD," October 30, 1897:
"I found the gun ('Altro') did wonderful execution, and up to 90 yards I have killed partridges single birds with the long brass cases quite dead. With the ordinary cartridge it is a most powerful weapon, and gives me the greatest satisfaction.—J. WADDINGTON-HUNT, Manor House, Daventry.

From the "FIELD," October 23, 1897:
"There are numerous situations where a gun of this sort ('Altro') may prove extremely useful. Having shot a great deal with this gun, I may perhaps claim to have arrived at a just estimate of its powers. It is certainly a most useful gun, by reason of the fact that with it one has a remarkably wide range of choice in the matter of loads."—HENRY SHARP.

NEW 40 PAGE CATALOGUE POST FREE.

"ALTRO" (Registered)
LONG-CHAMBERED 12-BORE, FIRING BOTH LONG AND SHORT CARTRIDGES.
THE BEST ALL-ROUND SHOT GUN FOR GAME AND WILDFOWL.

From the "FIELD," February 17, 1894:—
"I recently bought one of Messrs. J. and W. Tolley's 'Altro' 12-bore long-chambered Guns, and find it ALMOST AS EFFECTIVE AS AN 8-BORE."

We are now making these guns with shorter barrels adapted for the use of nitro powders. Sportsmen will therefore be able to shoot with guns of the some length as for game shooting.

PUNT GUNS.
Sir CHAS. ROSS, Bart., Balnagowan Castle, Ross-shire, writes:
"It may interest you to know I killed 52 geese in one shot to-day with your 1¾ in. Punt Gun. I believe this to be a record."

J. & W. TOLLEY, 59 New Bond Street, London.
(From 1 Conduit Street, W.), and
PIONEER WORKS, BIRMINGHAM.

J. & W. Tolley, Birmingham. Advertisement from *The Shooting Times*, 1898

---

forward allowance I found I had to make (even when using No. 1 shot, and with a gun I knew shot very hard), the range must have averaged 70 to 100 yards, and a friend of mine on the other side of the water, 150 yards away, told me afterwards that he occasionally expected to shoot a passing bird with his game gun just as it fell to my shot. I was using 2oz No. 1 shot, and with this charge the gun shot comfortably, though under 8lb; in fact, quite as easily as my 6½lb game gun shoots 1 1–16oz No. 6, and 32 grains E.C. No. 3.

Charles J. Heath.

Charles Heath was an eminent surgeon specialising in diseases of the ear. In relation to his professional life, his fame rested not merely on his published works, which were standard in their day, but also on a range of specialised surgical instruments which he designed to facilitate his speciality.

The French cases he refers to were almost certainly made by the Gevelot Company who had a history of involvement with the British trade beginning with the introduction of the breech-loading shotgun. Indeed, in the early days, they had the bulk of the British trade.

What probably worked against the wider adoption of thin, brass ammunition was the specialisation and dedication called for in terms of special wads, the tool to make the crimp closure and, above all, a gun, or certainly a set of barrels, adapted specially for them. In short, it was the system for the dyed-in-the-wool fowler, but not the casual dabbler.

With these thoughts in mind, it is interesting to move on to a wildfowling gun that was designed specifically to have wide appeal. The Tolley 'Altro' was promoted as being an all-round gun which would fire ordinary cartridges but still offer its user greater power on occasion. The most significant feature of the Altro was that it was chambered for the 3-inch paper-cased cartridge. This round had originated as a live-pigeon trap-shooting cartridge, the extra length being necessary to accommodate the potent charges favoured by some practitioners.

The potential of this powerful cartridge as a wildfowling round must have been obvious from the start and it is most likely that many gunmakers produced wildfowling guns adapted to it. However, the pre-eminence of Messrs Tolley as makers of wildfowling guns and their vigorous advertising ensured that their name became inextricably linked to the 'three-inch'.

The Altro gun was more a generic name than the designation of a specific model, indeed there were sidelock and boxlock hammerless guns as well as external hammer guns all marketed under the same title. Moreover, the same name was given to the maker's own cartridges.

What is interesting, particularly in view of the problems that had so recently occurred with the Lancaster 'Pygmie' 2-inch cartridge when fired in 2½-inch chambers, was the claim that all lengths of paper case could be used in the Altro. It had been found in the Pygmie affair that guns with abrupt forcing cones were more prone to permit the hot gases produced by the burning powder charge to fuse together a portion of the shot charge. This condition, called 'balling', results in a dangerous, solid missile capable of wounding bystanders who were well out of range of bird shot.

That no such problems were recorded with the Altro must mean that Tolleys had learned from the troubles that had dogged the Pygmie. The fact that the nominal 12-bore Altro was bored 'almost 10-bore' was probably the answer, for in this condition the forcing cone could only be a very small constriction.

In fact, the Tolley gun was a very versatile tool, well suited to the needs of the wildfowler who also had access to a spot of rough shooting. A 3-inch cartridge in the choke barrel is, after all, potent medicine for that wily, end-of-season cock pheasant who thinks he knows the range of a 12-bore shotgun.

J. & W. Tolley, Birmingham. Hammerless 12-bore Anson & Deeley 'Altro' gun

CHAPTER TEN
The Cartridge of the Heyday

'Up gets a guinea, bang goes a penny and down comes half a crown!' That was the wag's summing up of pheasant shooting. The guinea was claimed to be the cost of putting the bird over the guns, the half crown was its value dead and the penny was the cost of the cartridge used to shoot it.

In their different ways all these prices were remarkable. The cost of the live bird was probably inflated for the sake of a good line, having regard to the wages of an under-keeper, then about 16/- (80p) a week. The value of the dead bird was considerable in relation to the farm labourer's 14/- (70p) a week. The wonder is there was not more poaching. Most surprising of all was that good cartridges could be sold retail for just over 8/- (40p) a hundred. These were not the cheapest. At the very bottom of the scale were 'foreign cases, full charge, good wads, loaded with good English coarse black powder at 6/- [30p] a hundred', while at the top of the range were hand-loaded English cartridges with brass all the way up to the turnover at 13/- (65p) a hundred. So the penny cartridge was a sound, middle-of-the-range item.

To achieve such a price it was not necessary to do a special deal and buy in cart-load quantities. This was a very fair average of over-the-counter gun shop prices. Moreover, it was common practice for vendors to deliver free, orders of 1000 or more, by goods train to any station in the United Kingdom.

The reason for this keen price structure was simply the pressure of the market forces operating at the time. Huge quantities of cartridges were used although accurate figures are difficult to quote because military production of rifle ammunition is usually included in the figures. Significantly, there were three major British makers of cartridge cases, seven makers of smokeless powder, as many again making black, while up and down the country there were many makers of lead shot. Competing in the market were imported cases, components and loaded ammunition.

Such was the level of competition that, over the half century or so after the introduction of the breech-loader, prices of loaded ammunition had gradually decreased while, conversely, the quality of the product had risen. One further aspect of this story was the increasing quality of factory-loaded cartridges and because of the keen prices at which these were sold, small scale loading by gunmakers became increasingly uneconomic.

J. B. Warrilow, Chippenham. Advertisement from *Land & Water*, 1897

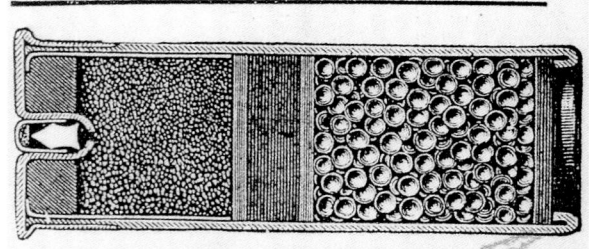

CARTRIDGES.
WARRILOW'S

NOTED CHIPPENHAM, Far Killing, with treble strong Powder, superior English Wads, Cases, and Chilled Shot, any size to order—Brown, 5s. 9d.; Blue, 6s.; Green, 6s. 6d.; Long Shot Reds, 7s. 6d. Unequalled Smokeless, reduced Recoil, Light Report, no smoke, but most effective, 7s. 6d.; Superior, 8s. 6d.; Best, 9s. 6d. Special for Pigeon Shooting, etc., 10s. 6d. These are special prices for Cash with order only. 500 sent Carriage Paid if Cash with order.—Get full particulars free from WARRILOW'S Gun and Cartridge Works, Chippenham, Wilts.

AGENT for KYNOCH'S, ELEY'S, and JOYCE'S SPECIALITIES; SCHULTZE, NORMAL, CANNONITE, BALLISTITE, and all the best powders. FIRST-GRADE GUNS at most reasonable prices. CARTRIDGE CASES, Wads, and Shot.—Apply to the general good man for all Sporting Appliances,

J. B. WARRILOW, Chippenham.

To understand the construction of the cartridge case at the turn of the century, we need to consider the problems faced by makers, above and beyond the basic function of sealing the breech of the gun for the fraction of a second that separated the ignition of the percussion cap from the exit of the shot from the muzzle. To ensure complete combustion of the propellant powder, the charge had to be restrained from moving for a fleeting instant and then released. Before that the cartridge had to withstand the vagaries of transport, whether to a home county or to one of the far corners of the Empire.

However, the greatest test for the cartridge was that it was expected to perform faultlessly in all sorts and conditions of guns and, more especially, in a variety of chambers. For, strange as it may seem, it was half a century after the introduction of the breech-loader before standard sizes for cartridges and guns to receive them were finally agreed and published.

The basis of the design and construction of the turn-of-the-century cartridge derived directly from a French patent obtained in 1846 by M. Houllier. In this specification *Monsieur* Houllier is credited with bringing together three features that had previously been used separately. These were a paper tube, a base wad or chaulk and a malleable metal capsule, which both held the

W. W. Greener, Birmingham. Cartridge advertisement from *Land & Water*, 1898

Eley Bros advertisement from *The Shooting Times*, 1898

other two components together and played a part in achieving the vital gas seal. Simple though these components are, fifty years' experience had taught the cartridge makers that they could not be casually swaged together. On the subject of the paper tube, Kynochs had the following comment to make in a magazine article published in 1898:

> We have ransacked England, Europe and America to find the very best paper for our purpose, and as a result, although there are hundreds of paper manufacturers, we have confined our purchases to practically two houses. To avoid splits, every sheet is examined and weighed separately and any showing the slightest defect are thrown out. In this way during the past year, we have passed over the scales separately, more than four million sheets of paper, and these have in a good light also received the most careful scrutiny and examination, and we believe that as far as the paper material is concerned, we have taken every precaution that is possible to avoid split tubes.

The same article went on to described how it was necessary to stabilise the water content of the paper to prevent the paper tube swelling. To this end, the paper was air dried in special lofts before it was rolled and the finished cases were stored for a further three months before use.

More evidence of the care with which the paper tube was made comes in the description of the flour paste used to stick it together. It was made with the best and sweetest flour available and tested by chemists on site before it was used.

Equally critical for the performance of the cartridge case were the characteristics of the brass head. This started life as a disc of brass punched out of sheet brass of high purity and accurate composition. In four stages, heated (that is annealed) between each one, this disc was drawn into the cup that became the head. The annealing temperature was critical for, if the brass was too soft at the instant of firing, the case would mould itself to the chamber and not contract afterwards. However, if too hard, it would split as the cartridge was fired.

One very vulnerable part of the cartridge case was its rim. About 1898 makers began to use an iron inner head, drawn to a shallower cup than the brass to strengthen the rim in better quality cases. These were sold as being 'Gas Tight'. The Gas Tight was but one of no less than ten grades of cartridge case which were offered in the Kynoch catalogue of 1902–3.

Something of the situation regarding the competing propellants on offer in the heyday is conveyed in the doggerel verse published in the April 1895 issue of *Arms & Explosives*. Entitled 'The Gunmaker's Lament', it is set to the air of 'The Sergeant's Solo and Chorus of Policemen' from the Gilbert & Sullivan opera 'Patience'.

> In gentle spring when shooting's not in season.
> Not in season,
> The Sportsmen study each powder anew
> Powder anew.
> Then swear by one of without rhyme or reason
> Rhyme or reason,
> And we gunmakers do some swearing too,
> Swearing too.
> We orders have for this and that and t'other
> That and t'other,
> But Law say that to stock all shan't be done,
> Shan't be done.
> Take one consideration with another,
> With another,
> Our lot indeed is not a happy one.
> Ah!
> Take one consideration with another,
> With another,
> Our lot indeed is not a happy one,
> Happy one.
>
> When not S.S., it's Schultze or it's E.C.,
> Or it's E.C.,
> New Smokeless Chilworth or the Amberite,
> Amberite.
> Some Cooppal, Cannonite or black it may be,
> Black it may be,
> Or Troisdorf, Walsrode or Ballistite,
> Ballistite.
> It's sure to be one if not the other,
> Not the other,
> For loading with all powders must be done,
> Must be done,
> Take one consideration with another,
> With another,
> Our lot indeed is not a happy one.
> Ah!
> When loading with all powders must be done,
> Must be done,
> Our lot indeed is not a happy one
> Happy one.

The profusion of competing powders was but one facet of the greater explosives industry which, in itself, derived from the upsurge of the chemical industry in the second half of the nineteenth

THE CARTRIDGE OF THE HEYDAY

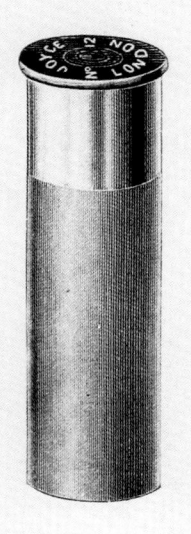

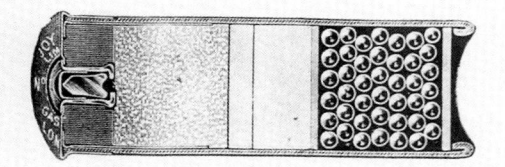

F. Joyce & Co. advertisement from *Great Guns*, 1896

Nobel advertisement for Ballistite

century. This came from the search for new, man-made compounds to replace or improve upon naturally-occurring substances as diverse as dyes, drugs and fertilisers. As the name suggests, Schultze was of German origin, which was to be expected given the prominent role played by the German chemical industry. In its advertising the Schultze Company of Gresham Street, London, claimed to be the vendors of 'The Original Smokeless Powder'. The first advertisements appeared in the sporting press in 1867 and the company was formed in 1869 to produce it in the United Kingdom. In fact, this claim overlooks the fact that in October 1865 there were advertisements for 'Prentice's Patent Gun Cotton Cartridges'. This reference from *The Field* is the earliest record so far found of a practical alternative propellant to black powder.

The basic feature of the production of Schultze powder was the treatment of small fragments of wood, essentially sawdust, with strong nitric acid to produce a chemical compound called 'nitro-cellulose'. Other cellulose bases, particularly cotton, could be treated in the same way to produce similar modified cellulose compounds and these formed the raw material of a whole class of shotgun powders.

Other organic materials can be treated with nitric acid to produce explosive compounds and the one of special significance at this point is glycerine, which becomes nitro-glycerine. In its pure state, this is an oily liquid which is very dangerous to handle as it will explode if subjected to a shock and, in its early days, it caused a series of accidents causing its use as an explosive to be banned. However, it was found possible to tame nitro-glycerine, famously to produce dynamite as a high explosive and Ballistite as a propellant for sporting ammunition. The latter, when used properly, was a perfectly safe powder, nevertheless the memory of the early problems with nitro-glycerine persisted and we see the claim 'contains no nitro-glycerine' much used by purveyors of various nitro-cellulose powders.

All these products needed to be presented to the user in a form in which they could be used safely. One way of doing this was to alter the density of the powder so that an appropriate charge could be measured with the same tools as had been used with traditional gun powder. These were what were called 'bulk powders' and were further described by the weight of the charge that occupied the same volume as the charge of black powder. Schultze, therefore, was made so that 42 grains, found to be the equal in power to 82 grains of black powder, occupied the same volume as the black powder.

Alternatively, for instance with Ballistite, no attempt was made to enable the old tools to be used and all that was specified was a charge weight. This resulted in the so-called 'condensed powders' and inevitably idiots used them with the old measures, wrecking guns as a result.

A root cause of many of the problems faced by cartridge loaders, and therefore the users of loaded cartridges, was that the new propellants were attempting to supplant the tolerant, age-old black powder. For such a powerful compound, gunpowder is remarkable in that, assuming it is kept dry and the appropriate grain size used, it is almost impossible to go wrong with black. What all concerned rapidly discovered was that a range of variables assumed vital significance when it came to loading the new propellants. Chief among these problems was the performance of the percussion caps and the pressure applied to the wads and powder as the cartridge was loaded. Some problems were more insidious, for instance, the water content of Schultze. It was found that perfectly loaded sporting cartridges charged with Schultze could be turned into virtual proof loads by driving off some of the water content of the powder. So the age-old practice of drying damp cartridges in front of the kitchen range had to become a thing of the past.

In the early years of the adoption of smokeless powders, it was found that certain percussion caps were better suited to the new propellants. For a while in the 1880s, there was a vogue for the larger-sized continental standard cap. Then, in a series of classic experiments, the results of which were published in 1893, William Dalrymple Borland, who was employed by the E.C. Powder Company, sought to build on prior experiments to bring some scientific order to the empiricism then reigning. Borland's method was to record on a photographic plate an image of the side view of a firing percussion cap. This, and additional experiments to measure both the heat and the pressure developed, enabled caps which would ignite the new powders reliably to be produced. As an incidental to his experiments, Mr Borland was able to demonstrate that a feeble blow or a blunt striker did not produce the best performance from the cap.

Once the powder had been successfully ignited, the resulting gases needed to be contained by an efficient obdurating wad to produce the best effect. The corps of knowledge concerning the wadding of the cartridge and its function derived directly from the long experience of using the muzzle-loading shotgun. Here a vast, almost comic, selection of materials had been tried. Favourite was a rolled ball of newspaper – *The Times* was reckoned to be the best. Discs were punched out of all sorts of material; leather, even slices of dried turnip were tried, along with specially made, thin metal cups.

Out of all this *ad hoc* experimentation, felt wads punched from old hats were ultimately found to be the best and the specially made wool felt wad seemed to combine the right meld of qualities of resilience and lateral expansion under pressure. Surprisingly, similar materials were inferior. Cow hair felt wads were not as good as wool felt. Worst of all was a wood fibre product, a sort of coarse thick blotting paper called 'Feltine'. In a trial reported in *The Kynoch Journal* in 1900, 'Feltine' was found to give results that were significantly inferior in velocity, density and evenness of pattern. Moreover, the fouling in the barrel after the trial was further evidence of the

The Derby Shot Tower. (PHOTOGRAPH COURTESY OF DERBY CITY COUNCIL MUSEUM & ART GALLERY)

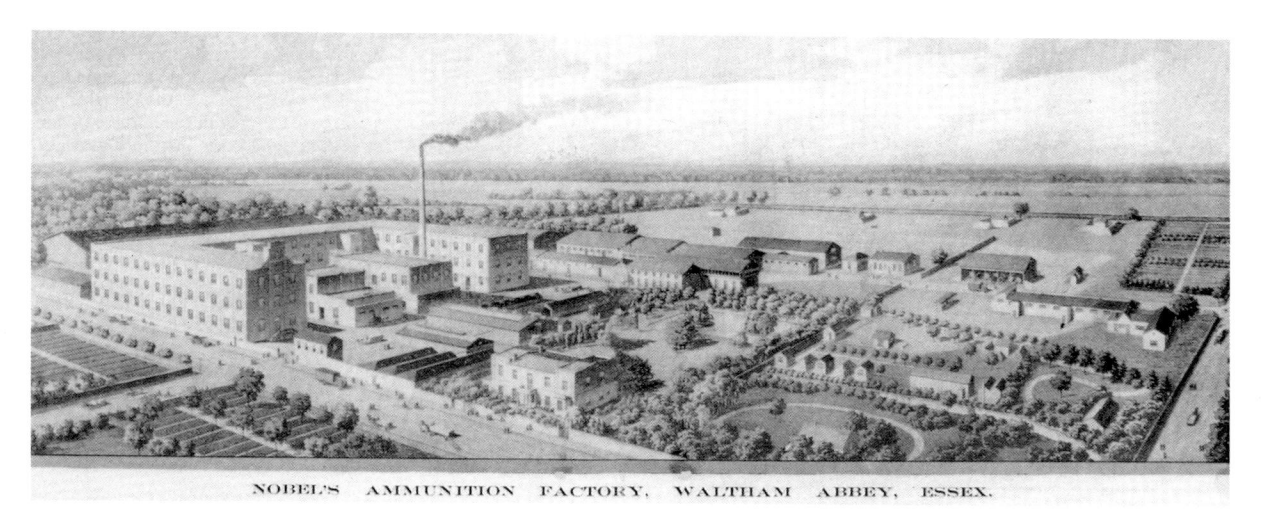

Nobel's ammunition factory, Waltham Abbey

inferior properties of cheap wads.

Like the wad that drove it from the barrel, lead shot was a product directly inherited from the days of muzzle-loading. Indeed the process by which molten lead could be converted into high quality shot simply by pouring it through a sort of colander and letting it fall from a height into a tub of water, had been discovered at the end of the eighteenth century. As remarked earlier, shot making was a competitive business, with the added complication that The Newcastle Chilled Shot Company based in Newcastle had its own size-grading standards, so it was necessary to specify the type required.

In addition, two basic qualities of shot, known as soft and chilled, were on the market. The soft shot was heavier pellet for pellet and this, along with its lower price, gained it adherents. The chilled was harder and so did not deform so readily in its tumultuous passage down the gun barrel. The result, especially in a choke-bored gun, was a better quality pattern, so chilled shot commanded a higher price in the market.

Exactly how this harder shot was produced was not entirely clear. Its name derived from the process of directing a blast of cold air through the falling molten lead. But some natural lead sources contained impurities such as arsenic and this had the effect of producing a harder alloy, which could be imitated by adding either arsenic or antimony to the molten lead.

The finished product, the cartridge for the sportsman to use in his gun, is all these diverse components brought together and properly loaded. There were essentially three routes by which this result could be achieved. The shooter himself could assemble his own ammunition, his gunmaker or another retailer could do it for him, or the retailer could buy ready-loaded cartridges from a factory. These three ways were not entirely separate, because a well-equipped, enthusiastic amateur might well have had the same, or even better, facilities than an ironmonger, who loaded very few cartridges. Equally, a gunmaker doing extensive business might have a set-up that approached the industrial.

The overall trend was towards the factory-loaded cartridge, for the simple reason that, for comparable quality, this was the cheapest option. The differential in retail prices was usually about 1/- (5p) a hundred, in other words a ten or twelve per cent reduction as compared with the gunmaker's own products. However, the picture was deliberately confused by the willingness of factories to produce special loads and cases. So, for an order of 25,000 cartridges, all of one type and taken within a year, the retailer could have a special printed tube of the colour of his choice.

These special loadings were but one more instance of a mentality that sought to preserve the myth of own manufacture, in the same way that small town 'gunmakers' tried to play down the existence of Birmingham as their source of supply, or the London trade was less than forthcoming about their out-workers.

The major factories used large, ingenious, multi-function tools derived from machines originally devised to produce rifle cartridges in the quantities demanded by huge armies equipped with breech-loading repeating rifles and rifle-calibre machine guns. On such a machine all the stages of loading are done without the interference of the operative.

At a lesser level of sophistication were hand-powered tools which batch loaded. The first such tool to gain wide usage was the Erskine, the invention of the Scottish gunmaker, James Erskine from Newton Stewart in Wigtownshire. This wooden tool, usually seen in the version that loads one hundred rounds per batch, was remarkably efficient. For instance, in a demonstration in 1866, the inventor produced fifty loaded pinfire cartridges in nine-and-a-half minutes. This, of course, was a demonstration sprint, the more usual way to work an Erskine loader was to have a team of men and boys feeding it. The wads were put up in a sort of rack holding the complete wad column and spaced so that they could be pushed, four at a time with a four-pronged rammer, into the cartridge cases. So, by having several wad boards with boys filling them, production could proceed apace, because it was a feature of the tool that the whole one hundred charges of either powder or shot were measured simultaneously then deposited in the cases held mouth upwards in the body of the tool. Likewise another hand would sit and turn over the ends of the cartridges with a treadle-driven tool, while yet another worker packed the final product.

The Erskine was a design of tool conceived in the days of black powder and, while it could be used for bulk nitros, doubts were expressed both about the accuracy of the charges thrown and the reliability of the wad seating. In view of the importance of both these factors, it is not surprising that in 1887 a tool was patented to answer these needs. Called 'The Climax', it was made by the firm of Dixons of Sheffield, very well known for their hand-loading tools, which incidentally were only one facet of a large business making cutlery and electro-plated goods.

In the Climax we see the retention of the idea of a block holding one hundred cases mouth-upwards, but with, in effect, three machines

Badge fitted to Erskine loader

Hundred cartridge size Erskine loader

through which the batch has to be passed in succession. In the first, the charge is thrown, ten cases at a time, by rotating an adjustable charge bar fed by a hopper mounted over it. Then the block is moved to the wadding tool and again ten cases are wadded at a time, but with plungers whose travel can be accurately adjusted. Finally, the shot is measured in a third machine, essentially similar to that which had measured the powder. The way to ensure accurate charges of both powder and shot was to keep an adequate supply in the hoppers and to turn the lever rotating the charge bar slowly enough to give the measures time to fill properly.

Much more simple tools were still sold for loading cartridges singly, but the range of tools on sale had greatly diminished over the preceding twenty years or so. Again the reason for the change was economic, the downward pressure on prices caused by competition and factory loading,

The loading room at Charles Lancaster, 151 New Bond Street, London. Notice the Dixon Climax loaders. H. A. A. Thorn was involved with the design of this machine. (PHOTOGRAPH COURTESY OF GEORGE YANNAGAS)

HEYDAY OF THE SHOTGUN

Some of the multitude of cartridges available in the heyday

resulted in a situation where only very small savings were to be made by refilling spent 2½-inch 12-bore cases. So, in reality, the tools on offer in the gunmakers' catalogues were a relic of the past, their greatest use being to produce special loads, light loads for ferreting rabbits, for boys learning to shoot, or perhaps blanks for an alarm gun.

So, by one means or another we have a loaded cartridge, ready to be taken off to the field in a locked cartridge bag or magazine (locked because cartridge pilfering was a persistent problem and whilst not to be condoned one must remember that 200 good cartridges cost more than a gamekeeper earned in a week).

In fact, the cartridge, or rather the range of cartridges on offer, gives us a very illuminating insight into just how palmy was the shotgunners' heyday. Consider the range of calibres that were readily available, all the way from No. 1s loaded with a pinch of dust shot, to punt gun cartridges containing a pound of what Col Hawker would have called 'glorious pills', that is mould shot of the largest size. If this size variation represents the length of the range, then the variety of cases on offer is its width.

The widest point of the range was the 2½-inch long 12-bore cartridge, of which there were about twenty-five different designs of case available in 1900. The choice was little short of bewildering, with subtle gradations from all-brass cases, via brass-covered paper with various lengths of brass, through different sorts of waterproof cases, ending at the bottom of the market with cheap and more or less nasty imported cases. In fact, it is difficult to give an exact figure for the number of case designs on offer, because some of the rival ranges would inevitably have been equivalents. The only true answer might derive from extensive dissection and even then the quality of the components would need analysis.

Moreover, the range on offer was not confined to the more popular sizes. For instance, pinfire

The controversial Lancaster 'Pygmie' cartridge, 2 inches long loaded with condensed powder and a full charge of shot. It was a dangerous failure because in guns with acute forcing cones, the shot was prone to ball

cases were available in all sizes between 4-bore and .360, including at least three different case designs in calibres as unusual as 24-bore. As for the centre fire, there was even more choice, with offerings as diverse as 3-inch 16-bore and a whole range, 4-bore to .410, of thin brass cases – Kynoch's 'Perfects'.

Added to this, these cases could be bought empty to load at home, or they could be loaded to order with one of the competing smokeless powders or one of the many brands of black powder. This was before the choice of wad, shot size, shot hardness or considerations like 'concentrators' which were card cylinders round the shot, in theory giving a denser pattern. Finally, there was the choice of a top wad, either an ordinary card or a 'Patent Breakable Top Wad', which, in theory, did not interfere with the developing pattern as the shot was ejected from the muzzle.

The options must have seemed endless. Arguably they were not all necessary, but they most certainly provide proof of the phenomenon I have called 'The Heyday of the Shotgun'.

CHAPTER ELEVEN
The Gun for the Future

The dawning of a new century gives a prompt to all those who delight in forecasting the future. Sitting behind their pints in the *Gunmakers Arms* in Bath Street, gunmaking prophets would inevitably look back on their own, and probably their fathers', memories if they sought to divine what lay before them.

The success of the gun trade in the nineteenth century which had given work to many, and comfortable lifestyles and fortunes to successful entrepreneurs, was based on a fortuitous combination of influences. Vitally, there was the popularity of shooting with all walks of society in the UK and in the Empire and the New World which gave unrivalled sporting opportunities, but the special ingredient that made the mixture fizz was technical innovation.

Given reasonable care and judicious repair, a soundly-made shotgun will last for at least two generations of sportsmen and, indeed, this had often been the way of things in the eighteenth century. Come the nineteenth century and percussion had succeeded flint ignition, breech-loading had replaced muzzle-loading and, for the last half of the century, roughly every ten years a new invention had rendered existing guns sufficiently obsolete to tempt a significant proportion of sportsmen to buy new guns.

At every stage of this evolution at least one pundit had gone into print to claim that perfection had at last been attained and, in fairness, given the convenience of say the centre fire cartridge or the rapidity of fire of the hammerless ejector, so it must have seemed. But each time the prophecy was proven false. It would have taken a very bold

---

Drawing from *Sporting Goods Review*, 1907, of the Hill & Smith hammerless gun with bar-in-wood action

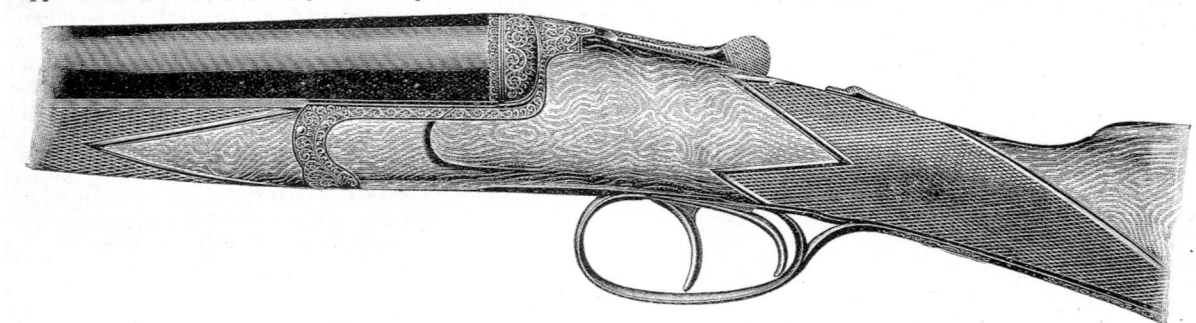

as the body has to be cut away to admit the plates, the metal is necessarily weakened at the angle where the upstanding breech face joins the horizontal part of the body. To strengthen it a greater thickness of metal could be introduced, but it was felt that the somewhat clumsy appearance so produced would spoil the appearance of a hammer gun. Mr. J. H. Walsh, the late editor of the *Field*, who devoted a very great amount of attention to the subject, said that although the term "break off" really belonged to the muzzle-loader, its retention was not unjustifiable as "it sometimes earns its name by breaking off." With hammerless guns,

There is no top fastening, but the body, although to all appearances so light, is heavily re-enforced at the points of greatest strain, the reduction in material and consequent saving in bulk and weight being effected by the removal of metal at the points least effected by the stress of firing. English gunmakers have been twitted with their belief that weight necessarily means strength. Here, while no good quality has been sacrificed for the sake of lightness, strength is secured by placing the metal where it will do most good.

The new gun is introduced as a revival of the skeleton bodied actions, or wood bar actions as they were called

prophet to say that there would be no more significant mechanical changes to the side-by-side double.

At the beginning of the twentieth century there were signs that the spring of invention that had sustained the prosperity of the nineteenth century was beginning to dry up. The last new invention, the single trigger, had not had the same universal acceptance as had the ejector before it.

There were some who believed that the next change would be more radical than anything seen since the introduction of the breech-loader. They would point to the shooting journals which were giving extensive coverage to the new efficient multi-shot Mauser self-loading pistol. In large measure, the difference between this new pistol and the revolver with which it was being compared, was not that it was self-loading but it was the application of rifle technology. It was the high-velocity jacketed bullet propelled by smokeless powder that made the huge difference to the shooting characteristics. Likewise, to make a comparable difference to a sporting shotgun would need revolutionary ammunition, which would certainly take longer to evolve and perfect. This long gestation period did not deter the visionaries and indeed Mr L.H. de Visme-Shaw went into print in the *Badminton Magazine* in 1902 with the following fascinating prediction.

> So in the case of the coming – we speculate only of course – change of form of the shotgun. When it first begins to make its way in the world people will call it a monstrosity and will say they would never be seen carrying such a thing, and will swear by the double hammerless against all blooming inventions whatsoever.
>
> A few years later the merits of the new type of gun will have brought it into general use; those who clung for a while to the double-barrelled gun will at length come to regard it in the same light as that in which they now view the muzzle-loader.
>
> The new gun, the shotgun which is to supersede our shotgun of today, will be a repeater. Fully loaded with its six or eight or ten cartridges, it will weigh appreciably less than the gun we now use; its barrel will be very short, not more than eighteen inches. One of the chief cries against the gun will be that it is unsportsman-like to carry such a weapon, its temporary critics driving wide distinction between the present-day practices of shooting with a brace of guns and the coming practice of adopting a repeater instead. Many a gunner of the past generation no doubt clung to the dear old slow and dirty muzzle-loader throughout his life, simply because he thought the use of the quick and simple breech-loader unsportsman-like. The repeater will be purely automatic; that is, the force of the recoil and its reaction will eject the fired cartridge, place the succeeding cartridge in the chamber, close the breech, and cock the hammer. Intercepting mechanism will ensure that the cartridge in the chamber cannot possibly be fired till the action has completed its work by securing the breech. This automatic ejecting, reloading the chamber, recocking the hammer and reclosing the breech, will be practically instantaneous, so rapid that however quickly the shooter wishes to put in a second shot, or a third, or a fourth, and so on up to the cartridge holding capacity of his weapon – he will invariably find the gun ready for him when he pulls the trigger.
>
> Before the repeating shotgun reaches its full development there must come about the perfecting of a powder in every way suited to it. Of this powder, but the smallest bulk – say a dozen grains – will impart sufficient velocity to the shot, it will be very much quicker than any powder now in use; while doing all of its work in eighteen inches or so of barrel; it will at the same time yield pattern and penetration as good as, perhaps a great deal better than, anything we can show today. Its pressure may be high, but this the build of the coming gun will obviate. The cartridges will be only about an inch and a half in length.
>
> There will be no change in the balance of the gun as shot follows shot. In the Winchester there is the objection of the alteration in balance; in the coming repeater there will be counter-poising mechanism which will keep the gun's balance constant from the firing of the first cartridge to the firing of the last.
>
> There will also be a mechanism enabling the shooter, by pressure of the left hand, instantly to convert his barrel from a cylinder to a modified or full choke and by release of the pressure to revert to a cylinder. How this may be effected I do not pretend even to hint, but it will be one of the chief features of the gun used in a generation or two, if not in our own. It is distinctly within the range of gunmaking possibilities.
>
> The barrel of the coming gun being no longer than about 18 inches, its stock, to equalise the poise, will be much less weight than our present day stock. Wood will have no part in its manufacture, it will be merely a band of blued steel, shaped to the correct outline, and flattened at the heel. Also it will fold back after the manner of the folding .410 of today thus enabling the sportsman to carry the gun – length about 22 inches, weight unloaded about four and a half

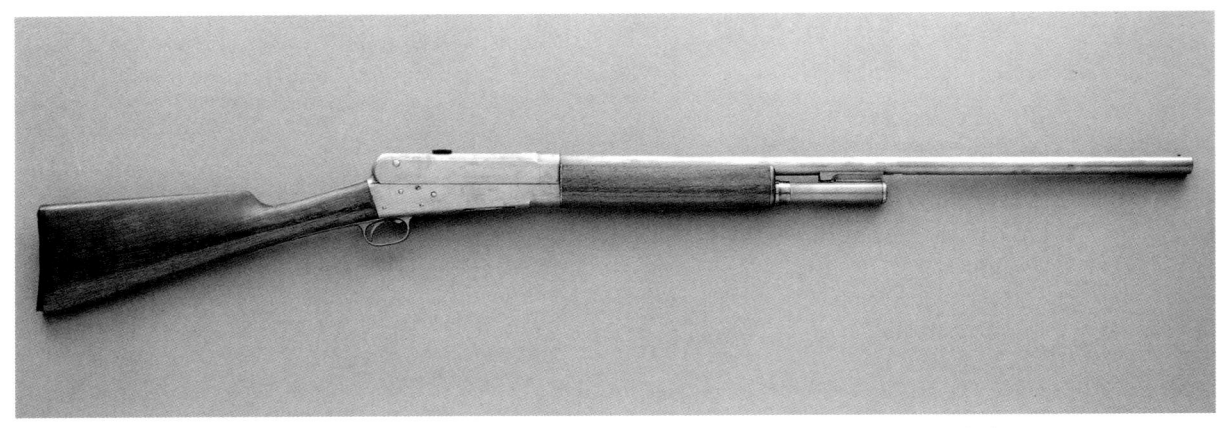

The prototype Browning 5-shot auto-loading shotgun. (PHOTOGRAPH COURTESY OF BROWNING ARMS CO.)

pounds – in the breast pocket of his shooting tunic. Light though the gun, excessive though, comparatively, the pressure of the powder used, recoil will be a thing practically unfelt even during the longest and busiest day, it being entirely absorbed by the spring work of the automatic action.

This prophecy was probably in part prompted by the appearance in 1900 of the five-shot Browning self-loading shotgun. Like the Mauser pistol, the mechanism used part of the energy of recoil to eject the spent case, replace it with a fresh cartridge and recock the lockwork. Unlike Mr de Visme-Shaw's visionary gun, the Browning was much heavier than the conventional side-by-side gun.

Instead of mentioning this problem, the commentators of the time regarded the Browning as a 'five-barrel gun with a single trigger mechanism', which is presumably a way of relating the new mechanism to guns with which they were familiar, such as the four-barrel Lancaster and the three-barrels of Boss and Green.

There was no perceived problem of using the Browning in place of a pair or trio of guns on driven game shoots, for it was considered that the new gun was merely a mechanical alternative to two or three double-barrelled guns and the necessary loaders. It was feared, however, that the retail price of £9 would make the gun available to the wildfowler or the rough shooter and so give him the facility to fire a third, fourth or even fifth shot in place of the two previously available and not only would more birds be killed from a flush, but, worse, more would be wounded by out-of-range shooting. Thus a serious depletion of wild stocks would result. By contrast, on preserved ground, the efficient rearing of birds would maintain the necessary stocks and, in any case, all the shooting was completely regulated.

There is absolutely no argument with the fact that, as a mechanism, the five-shot Browning self-loading shotgun is a masterpiece of mechanical ingenuity. Those who saw it as the way forward for the British sportsman were perhaps overly impressed by its technical features and, therefore, were not looking at the wider question of how such a gun would fit in with shooting as it was conducted in the British Isles.

Within the story of the nineteenth century, there were, after all, clear indications that the multi-shot repeater did not appeal to British tastes. The international nature of the gun trade in the nineteenth century is a factor frequently inadequately stressed, but it is certainly demonstrated by this facet of the story.

From the very first appearance of practical repeating pump-action shotguns in the United States in the 1870s, these guns had been available in Great Britain. The firm of John Rigby had imported what are believed to have been standard Spencer repeaters, while the firm of Charles Lancaster had imported Spencer actions which were retailed fitted with straight English-style stocks. The Lancaster/Spencer, as it was known, had the advantage of an extensive and favourable write-up and retailed at between £15 and £20 according to finish.

Later in the century Winchester repeaters, primarily the Model 97, were offered at £5. Despite the advantage of sound mechanisms, six shots in the magazine and prices that were on a par with the lower end of the market, all the indications are that there had been very little demand for such guns.

A difficulty with a pump-action repeater was that the user had to learn to pump between each shot and this problem was solved by the self-loading Browning. Those who believed that these clever pieces of machinery were the way ahead had overlooked the fact that part of the story of the sporting gun in the nineteenth century was its divergence in design from the arms carried by the soldier.

At the beginning of the century, the army musket was nothing more than a heavy, somewhat coarse, single-barrel, fowling piece. Indeed, if the opportunity presented itself, simply by being loaded with shot and wads instead of ball, the musket could be used on foraging expeditions. By the end of the century, the soldier had a manually-operated, repeating rifle, while the sportsman had a double-barrel, hammerless ejector shotgun. That is not to lose sight of the fact that they shared aspects of technology – breech-loading, steel barrels and smokeless powder – but, nevertheless, the two paths of evolution were steadily diverging and to promote the self-loading, single-barrel shotgun as the gun of the future would have been to assume that the paths of evolution would alter.

So, if the Browning or something derived from it was not to be the way forward for the sportsman on a formal shoot, this did not alter the fact that the gunmaker needed something new to attract his clients. A seductive claim was put forward by the advocates of the repeater – that because they had but a single barrel, they made for more accurate shooting.

Frederick Beesley, London. The Shotover Gun (patent date 1913)

Now absolutely nothing else succeeds like success, which is why gunmakers have always striven to make their guns lighter, handle better and shoot harder. For the same reason, gunmakers also discovered the commercial advantages of teaching their clients to shoot. In short, a more accurate gun is the gunmaking equivalent of the Holy Grail and the only practical way of giving a double-barrelled shotgun the sight picture of a single barrel is to mount one barrel on top of the other.

To put forward the idea that the repeating single barrel, or by derivation the double with superimposed barrels, was the only gun of the future was to overlook other clues, less clamorous than those of the self-loader, but there nevertheless.

Since it was obviously not such a new idea, the concept of a lighter gun might not have seemed like a pointer to the future. After all, the gun trade had always been able to make lighter guns, indeed there had been something of a limited fashion for shotguns weighing several pounds less than the standard back in the 1880s. The problem was, and indeed is, that the laws of ballistics cannot be ignored and the certain penalty for light weight is increased recoil if a standard cartridge is fired. It could be argued that these earlier light guns, by being too light, confused the message of 1900, when a range of technical improvements and changes in shooting practice were making a somewhat lighter gun a practical proposition.

The advent of what were called 'condensed' smokeless powders was a vital element in the new

Frederick Beesley, London. Locks from a Shotover gun. Note that one lock is effectively upside-down to deliver a more direct blow to the lower firing pin. (PHOTOGRAPH COURTESY OF FRED BULLER)

order of things. Simply because a charge of lighter weight was used, there was less matter to project from the muzzle and a reduction in recoil resulted. Additionally, the same nitro powder burned faster than black powder and so did not need the length of barrel. Moreover, the barrel itself could be made lighter for equivalent strength by using the improved steels which were becoming available.

In the shooting field there was the realisation that, for much driven game shooting, a somewhat lighter load than the old standard 1⅛oz was perfectly adequate. The slightly lighter loads could be just a little cheaper and, with a greater thickness of wadding, may well have been ballistically superior. In *Shooting on a Small Income*, Charles Walker reports on his own tests in which more even patterns were produced by 1oz of shot even in guns which had been regulated for 1⅛oz.

None of these changes were in themselves great but they were cumulative and taken together they meant that the standard gun could be somewhere between ½lb to 1lb lighter.

The question that needed to be resolved was how this change was to be used. The least radical change was that the gun for the twentieth century could be a shorter-barrelled, lighter version of the gun of the late nineteenth century, but there were persuasive arguments in favour of making better use of the new opportunities.

One possibility was a slight reduction in the calibre of the gun, while retaining the same weight both of actual gun and charge fired. This offered the chance to make a gun that was a degree trimmer, with the practical advantage that at the same time, the barrel wall thickness could be a little greater, so increasing resistance to dents. Alternatively, the metal in the barrel could be redistributed to give a slightly heavier breech, so enhancing the handling characteristics.

The most obvious calibres for the envisaged game gun of the future were the 14- and the 16-bores. Fourteen-bore cartridges had been on the British market from the beginnings of the breech-loading era. A good proportion of the pioneering Lancaster breech-loaders had been this size. While never as popular as the 12-bore, the Eley catalogue of 1899 still listed no less than five different qualities of cases of this size, varying from the cheap but serviceable brown quality to the brass-covered ejector and the all-brass 'Perfect'.

There were, perhaps, even more reasons to adopt the 16-bore, long favoured on the continent of Europe where its popularity is said to derive from a one-time edict of the French government that made this the maximum size permitted to civilians. The theory behind this law was that, should there be a rebellion and the sporting guns be used as muskets, they would be inferior to the larger bore muskets of the regular army.

However, far and away the most radical entry in the reduced calibre story was the introduction by Messrs Cogswell & Harrison of a new bore size – the 14¾-bore. The idea to split the small difference between 14 and 16 was nothing if not bold and no matter what the theoretical advantages of the idea, for the practical sportsman it must have raised the spectre of the bastard-sized special with all its problems and frustrations. So it is all the more surprising that this idea was followed up and guns and cartridges were actually produced in this new size.

If this strange size was considered radical, it paled beside another visionary scheme, the so-called 'Vena Contracta' which, as its name suggests, embodied the notion of constricting the bore of the gun. It was found by experiment that, contrary to reasonable expectations, it was possible to fire a contemporary shot cartridge with card and felt wads safely through a barrel several sizes smaller than the size of the over-powder wad. Equally surprising was the proven fact that

Cogswell & Harrison 14¾-bore cartridge

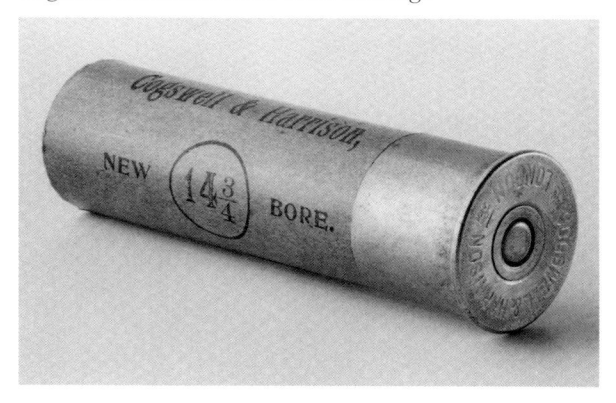

COGSWELL & HARRISON, Limited

New $14\frac{3}{4}$-bore Ejector Gun

		CASH	CREDIT
Best quality materials workmanship and finish	}	70 guineas	84 guineas

Conditions of Sale, see page 14.

Best quality materials workmanship and finish	}	70 guineas	84 guineas

Conditions of Sale, see page 19.

IN introducing this new Gun a comparison is made with a modern 12 bore.

Like a 12 bore:

> Same charge of shot, namely $1\frac{1}{16}$ or 1 oz
> Equal patterns, both inside and outside of 30-inch circle.

Improvement on a 12 bore:

> Higher velocity.
> Greater penetration.
> Longer range.
> Six ounces lighter.

The ordinary 12-bore cartridge of to-day is the same in general dimensions as it was 50 years ago, it was designed to shoot 3 drs. or $3\frac{1}{4}$ drs. of black powder and $1\frac{1}{4}$ or rather more of shot.

The new $14\frac{1}{4}$-bore shoots Cogswell & Harrison's smokeless powder "Vicmite" which occupies a reduced space in the cartridge and burns without residue.

141, New Bond Street, *and* 226, Strand, LONDON

Advertisement for the Cogswell & Harrison 14¾-bore gun and cartridge

all the parameters – velocity, pressure and pattern – were normal. The patentee in 1893 of this odd notion was Horatio Phillips who, among other things, was part of the editorial staff of *The Field*, which explains, of course, why the gun was so well written up in that journal. His patent covered the idea of constricting a 12-bore barrel down to 18-bore at about a third of the way up the barrel.

The coverage in *The Field* of this gun's performance gives some more of the facts surrounding it. Its main retailer was the firm of Joseph Lang in London's New Bond Street, but it emerged that the guns were, in fact, made by Webleys in Birmingham. From the same source came the information that another London retailer was Frederick Beesley in St James's Street. However, perhaps most surprising of all, H.A.A. Thorn, 'Charles Lancaster', wrote in to say that he had built a 3-inch chambered 12-bore pigeon gun which ended up as a 16-bore. This was condemned as a failure on account of the high chamber pressures that resulted from this set-up.

Amid these stirrings towards a new calibre shotgun for the game shot, there were others that demonstrated an upsurge in interest in an ultra-small-bore shotgun, the .410. The origins of this calibre are somewhat obscure but the most likely explanation is that originally it was simply the continental 12mm cartridge translated to imperial measurements. Until the 1890s, it was only listed in the cartridge makers' catalogues as a pistol and rifle cartridge. In this form the longest case listed is the 2-inch version. The first appearance of a shotgun cartridge of this calibre is in the Eley catalogue of 1895. However, the production of factory-loaded rounds suggests that a demand had been created, presumably by gunmakers hand loading shot rounds in rifle cartridge cases.

To fire this round, the trade offered a shotgun version of the centre hammer, side- or under-lever action, also employed as a rook rifle. As always, the readers' letters of the shooting papers offer an insight into what was happening out in the coverts and in 1908 a correspondent wrote to *The Field* to enquire as to the effectiveness of the .410. This drew a reply from H.A.A. Thorn, proprietor of the firm of Charles Lancaster, who was ever alert to the potential for free advertising offered by the correspondence columns of the sporting press:

Henry Thorn's son, Alan, with .410 gun.
(PHOTOGRAPH COURTESY OF MRS A. THORN, FAMILY ARCHIVES)

> I prepared a special .410 for my small boy's use in 1895 with which he was able to kill rabbits in a very satisfactory way. I followed this up by always copying the gun and recommending it for youngsters' use when learning to shoot.

That others agreed with these ideas is confirmed by the fact that when B.S.A. introduced their bolt-action, single-shot gun, they made it not just for the standard .410 but for the more potent 2½-inch long cartridge, which was originally developed for this gun.

Returning to the central theme of this chapter, the gun trade's need to develop their products and

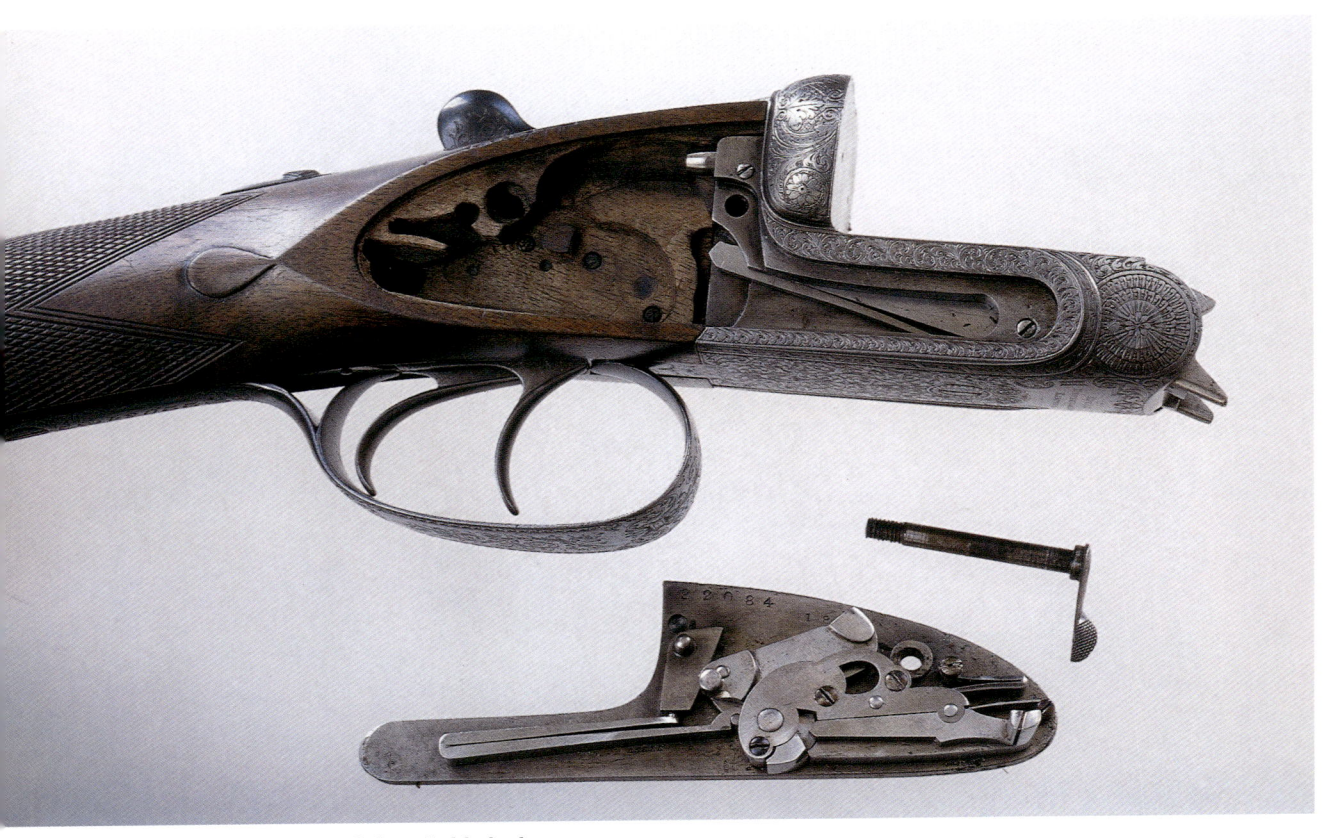

Holland & Holland hand-detachable lock

extend their market, one of the problems facing the trade at the turn of the century was that mechanical evolution concentrated on design to the exclusion of individuality among the upper echelons. One way to restore some degree of house-style was to explore new styles of engraving. If this, at the same time, made new guns look modern, it might also have the added effect of persuading a client to buy a new gun rather than look for a second-hand bargain. After all, the new season's look is a proven formula that sustains any trade.

Historically in North America, the rest of Europe and Asia, there have been many ways of ornamenting guns. Woodwork has been carved and metal chiselled and there have been inlays in all sorts of materials all over the gun. In the face of all this, the British shooter has remained remarkably conservative in his choice of gun decoration. Exponents of new styles of engraving would have known that they would be swimming against a strong current. Another difficulty facing the would-be designers was that the traditional scroll and bouquet motifs are not simply decoration applied like wallpaper to a blank wall. The engraving on a gun should enhance its lines while, at the same time, disguising such mundane things as joints. However, given the problems, the effort was worth making, adding another facet to an already complex story.

A topic certain to raise the temperature of the debate in the *Gunmakers Arms* would have been the question of foreign imports and the problems of exporting. Indeed, one way or another, there was a persuasive case to be made for the notion that foreigners would capture the whole of the lower end of the gun market, not just the home market, but exports as well.

That wages could be reduced to Belgian levels was unthinkable, the only other alternative was mechanisation. This was nothing new to Birmingham. Bonehills had a period, in the early

THE GUN FOR THE FUTURE

A typical turn-of-the-century .410 shotgun

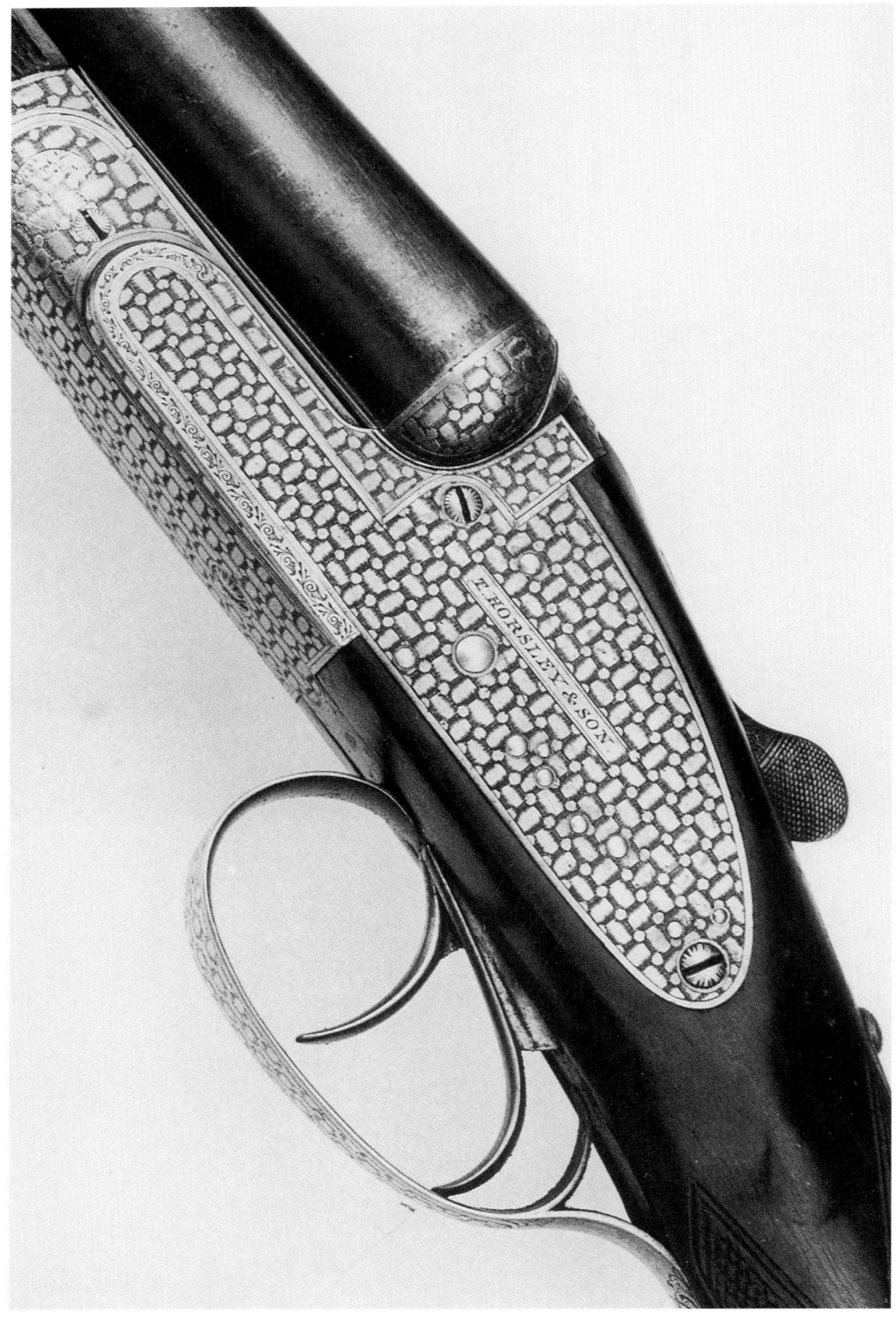

Thomas Horsley of York. Unusual pattern engraving

THE GUN FOR THE FUTURE

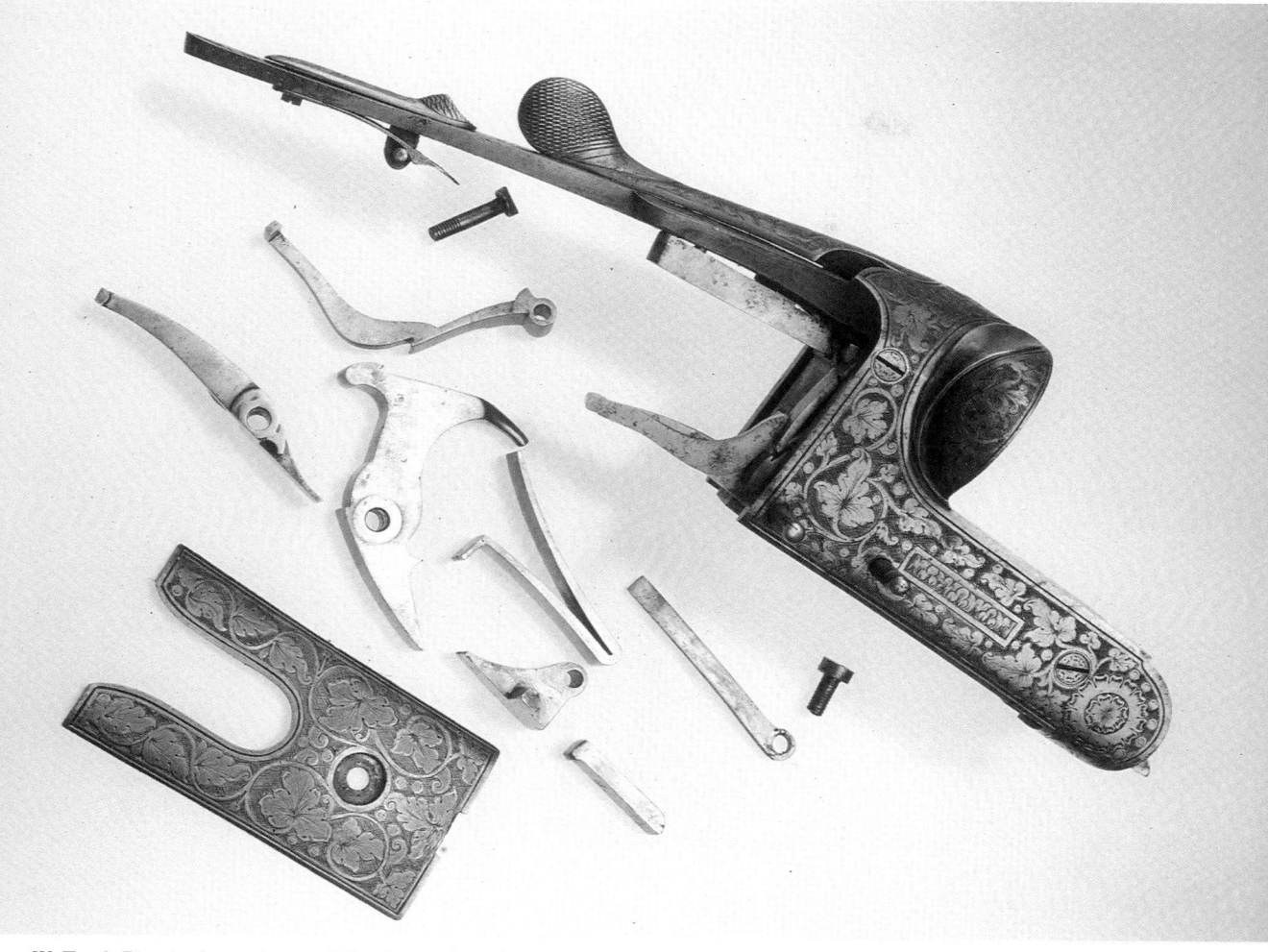

W. Ford, Birmingham. Anson & Deeley-actioned gun with unusual engraving

1890s, when their machine-made 'Interchangeable' guns had done well until the American tariffs had been raised and a vital market lost. Insufficient markets meant insufficient profits to fund the new machinery or to create the will to progress, so the enterprise had withered away.

The more perceptive of those who considered the prospects for the Birmingham gun trade in the new century, realised that the answer for the mass market lay with their political masters. For it was only the politicians who could ensure a home market and persuade their counterparts in potential markets to open them to British goods. The best hope was the Empire, providing these markets could be preserved. If this could be done, then Birmingham could make single-barrels to compete with the Harrington & Richardson and a new double-barrel, designed for machine production to retain markets and sustain the gun trade. If this were not done, then the age-old pattern of workers taking their skills to other industries would continue and one of the once-great industries of Birmingham would fade into history.

Epilogue

In 1900 nobody could have predicted how the century would unfold. With hindsight, we know the awful facts. In practically every village, town and city in Great Britain, there stands a memorial to those who died in The Great War. While the total casualty figure is impossible to comprehend, the local lists with their groups of brothers and cousins, along with those who were wounded, when related to the size of the community, brings home the proportion of women who became widows and the girls who were condemned to spinsterhood. Then we begin to understand the human cost that was the legacy of the 'war to end all wars'. Add to this the emotional scars and altered perceptions of those who survived the trenches, the north Atlantic convoys or the Royal Flying Corps and we realise how true was the simple statement, 'Things would never be the same again'.

By the end of the war, sporting gun production had virtually ceased as firms with any metal-working background had been swept up into the war effort. Cartridge production had been severely curtailed, while out in the countryside, keepers had gone off to war and vermin had multiplied. Even the face of the landscape had changed; coverts were felled for timber as land went under the plough to attempt to feed a population faced with starvation by the U-Boat blockade.

There were also positive outcomes of the war; the advances in technology, especially the motor car. On top of this came The Great Depression that followed the stock market crash of 1929.

Inevitably, the challenges facing the gun trade had been altered by the war. The need to innovate had been sharpened. Think of all the second-hand guns thrown on to the market by the casualty lists. Perhaps this was the reason the light-weight gun developed not as a smaller calibre but as a light 12-bore. By the 1930s, all the top-flight gunmakers were offering guns that weighed close to six pounds, even some a little under. Indeed, the trend went even further in a somewhat surprising direction; the two-inch long 12-bore case was revived, now loaded with a charge previously associated with a 20-bore. This gun was promoted as 'The Twentieth Century Gun'.

This craze for light-weights is probably one of the reasons that, in the aftermath of the war, the over and under gun did not sweep all before it (as might have been expected by the rash of patents that preceded the conflict), as it is much more difficult to produce a very light over and under. The 'vertical barrel' gun was further hampered by the fact that the British trade concentrated on the top end of the market. Moreover, there was a degree of prejudice against the new layout. Both these factors were to change with the more affordable Belgian Browning of the 1930s and the popularity this gun enjoyed with the clay pigeon shooting fraternity.

The most obvious legacy of the war was on the lower end of the market, where firms sought to apply the experience they had gained in the mass-production of precision metal-work to peace-time projects. This trend was to be seen in a whole range of products in Europe and the United States, but it is relevant to this story in the appearance of the Ward 'Target' and the B.S.A., both side-by-side, double-barrel shotguns. If anything, the Ward was the more innovative with its extensive use of coil springs. By comparison the B.S.A. was much more prosaic, being a re-worked Anson and Deeley. Very high hopes were held for the new B.S.A. but, despite valiant efforts, familiar old problems, especially cheap, imported guns during the difficult years of the depression, conspired against it.

These forces might have been expected to

favour the multi-shot, repeating gun but they never became widely used. The consensus remained that such guns were both unsporting because of their enhanced firepower, and unsafe because they could not be immobilised by being carried open like a double-barrel.

The mass-production techniques learned during the war and the hard economic times of the 1930s both favoured the machine-made single-barrel. Not surprisingly a gun of this type was one of the 'peace products' which the Vickers firms turned to in an attempt to make some use of factories that had made machine-guns and aircraft during the conflict.

By comparison, the Second World War had a less radical effect on the gun trade, because the damage had already been done by the First. The extensive destruction of property was accompanied by the loss of gunmakers' records. Elsewhere, the decline continued with old firms fading from the scene and not being replaced by new.

Two overriding problems faced the gun trade in the second half of the twentieth century. First, there simply were no new inventions to make existing guns obsolete. Second, such was the quality of the guns made during the heyday that they continued to serve the grandsons of their original purchasers.

What lies in store for the future? It could be that the adoption of shot less efficient than lead could give rise to a new breed of shotgun, but as Mr. de Visme Shaw proved, predicting the future is a risky business.

Appendix One

A Gunmaker's Catalogue

The bicycling boom that gave mobility to the masses in the last decades of the nineteenth century brought with it a welcome boost to the trade of the provincial gunmaker. Gun making, of all the trades that were then represented in even the smallest market town, benefited because the gunmaker, alone among the metal workers, had the tools and the skilled workmen able to work to the scale and precision needed by the new machines. So, all over the country, established gunmakers competed with specialist cycle dealers to retail and build bicycles under their own brand name.

These own-brand cycles were, in the main assembled from parts bought in Birmingham and so the established pattern of the gun trade was maintained. Moreover, many gun-making firms in Birmingham also diversified into bicycles. Mention can be found in various sources of bicycles made by Tranters, Enos James, J. & W. Tolley & Co., Holloway and Kynoch. This trend spread on to London where Cogswell & Harrison began making bicycles in their Gillingham Street factory.

To publicise their wares, a stratagem adopted by retailers up and down the country was to tote the records held by their machines. Since nobody had previously officially timed travel by bicycle, it was possible to claim record times for journeys between every town and village in the land. In this way an active young man could become a 'champion' and his sponsor could have a glut of advertising copy.

As but one small part of this overall scene, the firm of Hillsden & Sons, cycle and gun manufacturers of Folkestone in 1894, opened a branch establishment in Dover and William John George, 'The Champion of Kent', was employed as manager. Four years later Mr George emerged as the new owner of the business and issued a wide ranging catalogue which provided a priced snapshot of the provincial trade of the time.

However, this document needs to be approached with more than a touch of caution, right from the front cover. Notice the Royal Arms displayed in the top left-hand corner. These were certainly something Mr George had no right to. The claim to be a 'Government Contractor', if others of this ilk are a guide, refers to some trifling sale to the garrison down the road.

Underneath the Royal Arms is an impressive display of medals such as were awarded at exhibitions, but since there is no other mention of these 'awards', they can only be regarded as spurious window dressing.

Turning to the contents of the catalogue, it will be noted that there is no visible name on any of the guns on offer. The most likely explanation for this is that the blocks used to print this catalogue were loaned by the suppliers of the guns. Thus, if a sufficiently careful search were to be made of contemporary catalogues and other advertising, the same illustrations would be found in other contexts.

This leads us to the question of the true maker of the guns offered. Some are instantly recognisable as the products of Liege, the rest, despite hints designed to imply local manufacture, were almost certainly obtained in Birmingham. Logic tells us that if there really was a factory in Dover, much more would have been made of it in the catalogue. Details and invitations to visit would have been prominently displayed.

Despite all these caveats, the W.J. George catalogue is a fascinating document. The prices, the range of goods freely sold and the social insights offered are all worthy of study and shed further light on the scene in 1900.

GOVERNMENT CONTRACTOR.

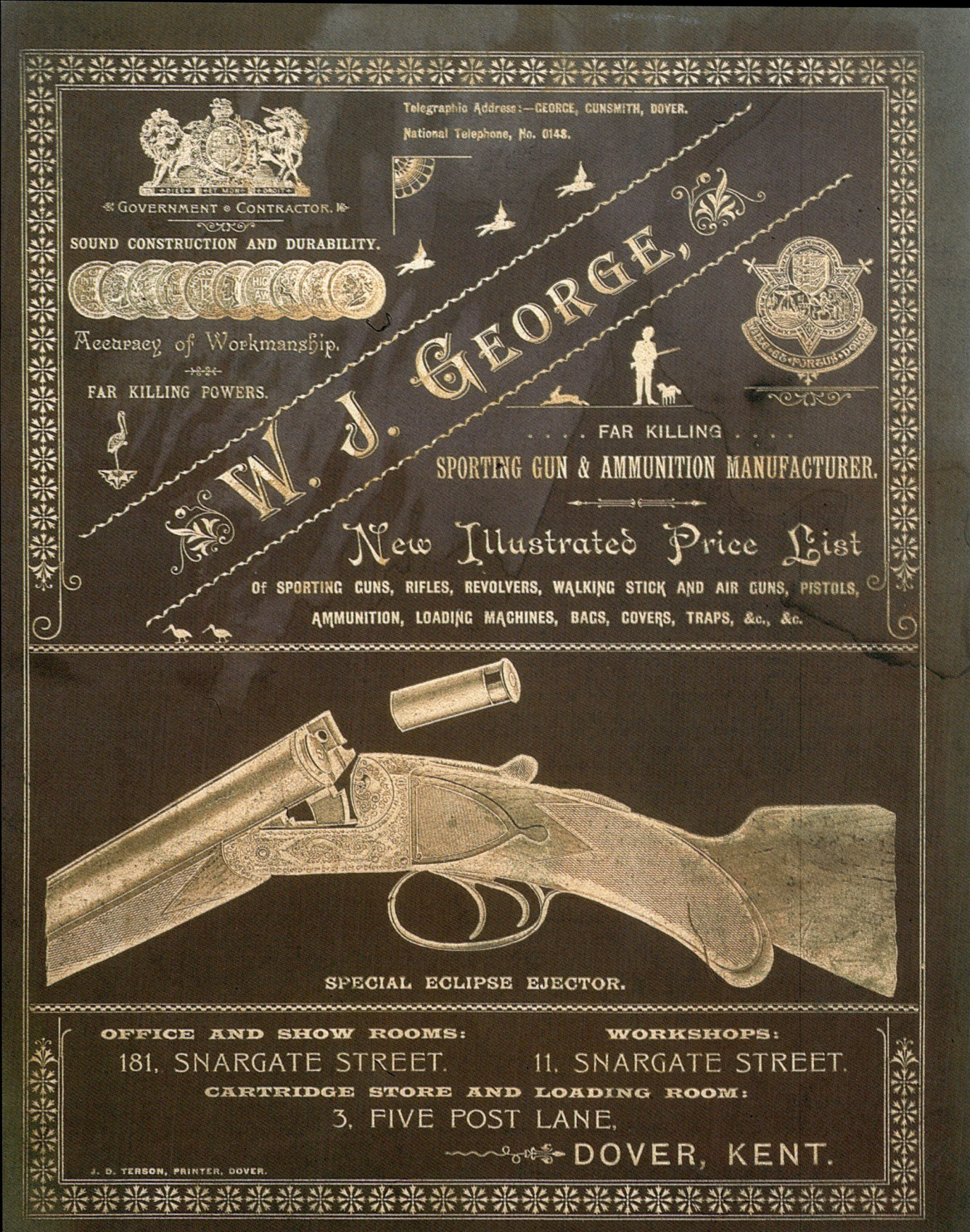

SOUND CONSTRUCTION AND DURABILITY.

Accuracy of Workmanship.

FAR KILLING POWERS.

Telegraphic Address:— GEORGE, GUNSMITH, DOVER.
National Telephone, No. 0148.

W. J. GEORGE,

.... FAR KILLING ....

SPORTING GUN & AMMUNITION MANUFACTURER.

New Illustrated Price List

Of SPORTING GUNS, RIFLES, REVOLVERS, WALKING STICK AND AIR GUNS, PISTOLS, AMMUNITION, LOADING MACHINES, BAGS, COVERS, TRAPS, &c., &c.

SPECIAL ECLIPSE EJECTOR.

OFFICE AND SHOW ROOMS:
181, SNARGATE STREET.

WORKSHOPS:
11, SNARGATE STREET.

CARTRIDGE STORE AND LOADING ROOM:
3, FIVE POST LANE,

DOVER, KENT.

J. D. TERSON, PRINTER, DOVER.

ALL PREVIOUS LISTS CANCELLED.

TERMS OF BUSINESS.

My Terms are **CASH WITH ORDER,** or against Invoice, and these terms will be strictly adhered to. Buyers must pardon me for being so rigid, but the prices are calculated for Cash, without any risk of loss, and doing business for ready money at the lowest possible price, I cannot make exceptions by altering or concessions—deviations being unfair to other buyers, and I try to act equitably towards all.

Remittances may be made by Post Office Orders, Postal Orders, Registered Letters, Cheques, or Stamps. Cheques, etc, for safety, should be crossed to my Bankers, Lloyds, Limited, Dover.

Any Article that does not give entire satisfaction will be willingly exchanged at any time.

All Guns are carefully packed. Carriage is paid by me (unless otherwise specified) and all Goods except Ammunition are sent by Parcels Post if possible, if not, per Passenger Train to nearest Station,

Foreign Orders must be accompanied by remittance, or an order upon an English Bank or Mercantile Firm for payment against Bills of Lading. Guns for Calcutta, or any African or Australian Port should be enclosed in a zinc-lined case, for which the charge is for One Gun 4s., and every Additional Gun 1s.

On all Orders for Guns built to order, a deposit of not less than 20 per cent. (4s. in the pound) will be required.

Full market value allowed for Guns taken in part exchange for new. All arms sent to us for repairs or exchange should bear the sender's name and address.

Repairs or Goods for Exchange in all cases must be sent Carriage Paid.

I guarantee all my Guns and Rifles to bear the Legal Proof Marks as instituted by the Authorities, therefore I do not accept any further responsibility. The Illustrations herein shown are intended as a general guide, and are not considered binding as to details

All Barrels described in this List as Damascus, Curly Demascus, or Fine Curly Damascus, are machine forged, made on the Improved Economic Compound Coiling Process. All Twist Barrels are of the Best Skelp, Machine Forged. Any arm willingly exchanged if not approved of.

My Guns are Known throughout the World. See that every Gun bears my Name.

I hold one of the Largest and most Complete Stock of Sporting Guns and Rifles. If you do not see what you require in this List, please write, particulars shall be sent fully and at length.

Lowest Quotation for High-Class Sporting Goods of every description. Lady's and Gentlemen's Cycles and Accessories per return of post.

All Orders, Letters, and Correspondence to be addressed to W. J. GEORGE, Gunmaker, Dover, Kent.

All Parcels sent to me, Exchanges, or Guns, should bear sender's Name and Address, and should be sent Carriage Paid (otherwise they will not be accepted), and instructions sent by same post.

Being the Dover Representative for Messrs. HUMBER & Co's. Celebrated CYCLES, &c., I shall be pleased to quote Lowest Cash Prices on application. I also manufacture the "Dover" Cycle.

REPAIRS, CONVERSIONS, &c.

Re-stocking Guns, from 16/6; re-blueing and re-hardening, 4/- to 12/-; re-browning barrels, twist or steel, 4/6; Damascus, 5/6 to 8/6; top-lever spring, 3/6; repairing, overhauling, cleaning locks, polishing, and re-browning barrels, and making as new, 20/-; new hammers, 2/6 to 4/6; new striker and spring, 2/-; new main spring, 2/6; new fore-end, spring, or bolt, 12/6; choke boring old barrels, 3/- to 6/- per barrel; lengthening stocks, 5/-; taking dents or bruises out of barrels, 2/6; tightening up shakey actions, from 7/6; chambering for perfect cases, 4/-; converting pin to central fire, 25/-; fitting new barrels, fine twist, from 35/-, steel, from 50/-; curly Damascus, from 50/-. I have a Special Department for Repairs, but owing to the peculiar nature of most repairs, it is impossible to adhere to a set of standard prices, but I shall in all eases, where practicable, keep to the above prices. **The above Prices include Return Carriage.**

Illustrations are of Standard Patterns, and made as follows:—bend of stocks, 2-in. to $2\frac{1}{2}$-in; length of stocks, 14-in. to $14\frac{1}{4}$-in.; length of barrels for 12, 16, and 20 guage, 30-in.; 10 guage, 32-in.; and 8 guage, 34-in. to 36in. Locks with hammers on them are all **rebounding.**

Customers requiring Special Measurements must state the same on their **Orders,** also if **straight,** half-pistol, or full-capped hand be required (unless the same is shown in illustration).

Any Gun, &c., not enumerated in this List will be specially quoted for on receipt of specification.

ALL GUNS showing Pistol Stock, Circular Hammers, or Solid Strikers may be had with STRAIGHT STOCK, Round-body Hammers, or Nipples or Plungers at same prices.

Telegraphic Address:—GEORGE, GUNSMITH, DOVER.   Telephone No. 323.

INTRODUCTION.

AFTER YEARS OF SUCCESS.

Read what the "Dover Standard" says:—

"W. J. GEORGE Cycle and Gun Manufacturer, Snargate Street.—Now that the Shooting Season is on call and select your Gun. He stocks all the best makes, besides ammunition of all kinds. Cycles by all the best makers can be obtained here, besides Lamps, Bells, Spanners, and all other Accessories. Repairs are executed on the premises, and riding taught by experienced instructors. Mr. W. J. GEORGE, the Champion of Kent, is a mine of information about Guns, Cycles, and Accessories."—*Extract from "Dover Standard," December 19th, 1896.*

DEAR SIR,—In issuing this Annual Catalogue, I wish to thank Clients once more, for the many warm and encouraging expressions of approval, with practical hints and suggestions, many of which have proved of value to me. During the past season I have had an enormous increase of business, (and this I take it, is the best possible indication that my business is conducted on sound business principles, for the public are sure to support those who provide the best value for the least outlay.) My policy, impressed on my employés, being to put ourselves in the buyer's place, do for him as we would be done by, and never misrepresent or overcharge. Unfortunately no system can prevent errors owing to the neglect of those who administer it, but when an error occurs please believe it accidental and not intentional; and I shall esteem it a favour, in case of the slightest neglect on the part of an employé, if you will write a private letter to me, and I will stop its recurrence, as my desire is that all should receive that attention and civility which will make it agreeable as well as profitable to send to my establishment in preference to others.

The substantial development of business in all departments is such that with the enlarged workshops etc., at my disposal, I now hope to be able to execute all orders from stocked goods immediately.

For the coming season I have not found it necessary to introduce many changes, my aim is to turn out Guns more with a view to utility, strength, sound construction, and durability, than to elaborate finish and outside show, and it is these solid qualities together with their extraordinary long distance killing power, which has brought my Guns to the front throughout the country, Most careful attention is paid to the construction and shooting of every Gun, and I would strongly advise all good shots, or those who would wish to become good shots, to have their Guns built especially to their requirements, as the fit of a Gun is of the utmost importance. For this purpose you will find instructions in Catalogue how to give exact measurements of Stock, etc., required. Many a would-be sportsman has after sundry trials cast his gun aside and given himself up for a complete failure as a shot, under the belief that his eyesight was at fault, whereas the fault was in the fit of the Gun. Some of my customers ask how it is my prices are so much below the prices of the ordinary Gun Makers? The reasons why my prices are so much below those of other Gun Makers are several, viz.:—my terms are *Cash only*, I therefore make no bad debts, and employ no travellers, I do not spend hundreds of pounds yearly in advertising, although I do not under-estimate the advertiser's pen, but the value of an advertisement depends on the goods behind it—and I would here respectfully warn buyers not to be led astray by glowing and expensive advertisements. My Guns are far in advance of any other placed upon the market at the price, on the score of sound construction, durability, and far killing powers. I consider a Gun is an article that cannot be made too good. I am sensible of the fact that there are some Makers or Dealers advertising Guns at lower prices than me, but I can safely go to this extent, and say you cannot buy a Gun of the same quality, same finish, same improvements, at the same price as I sell mine. To those requiring such cheap articles, I would call their attention to the page in this Catalogue, where they will find mention made of (I don't say cheap, but) common made shoddy, manufactured by certain firms in Birmingham and Belgium, whose only object is to throw the Guns together at the lowest possible cost. It is however gratifying to observe that Sportsmen are beginning to appreciate the difference between a jerry-built Gun and a Gun built for strength, durability, and far-killing,—the position to which my Guns have attained,

My Customers may rest fully assured that no effort will be wanting on my part to still further develope by the closest personal attention, the utmost courtesy, civility, and readiness to serve, and the continual putting in practice of those most popular principals of business which have led to success in the past.

To those who have not given me a trial I would say "Do so," you will not regret it; and to those who have, I again tender my warmest thanks, and feel certain that if I lose their confidence it will be my own fault, and not the fault of Goods supplied to them by me.

I remain, yours faithfully,

W. J. GEORGE.

N.B.—I guarantee all Guns and Rifles to bear the Legal Proof Marks as instituted by the Authorities. therefore I do not accept any further responsibility. The Illustrations herein shown are intended as a general guide, and are not considered binding as to detail.

THE EXCELSIOR GUN, 50s.

(WITHIN REACH OF ALL, AND GUARANTEED ENGLISH MAKE).

I have noticed there is a great demand for a cheap reliable Double Breech Gun, and have therefore introduced the following Two New Models into this Season's Catalogue; the vital parts, namely locks and body in all wearing parts, have had special attention in fitting; the barrels also have special attention in the boring of same to ensure a good pattern and great penetration, combined with far-killing powers, by my special boring process.

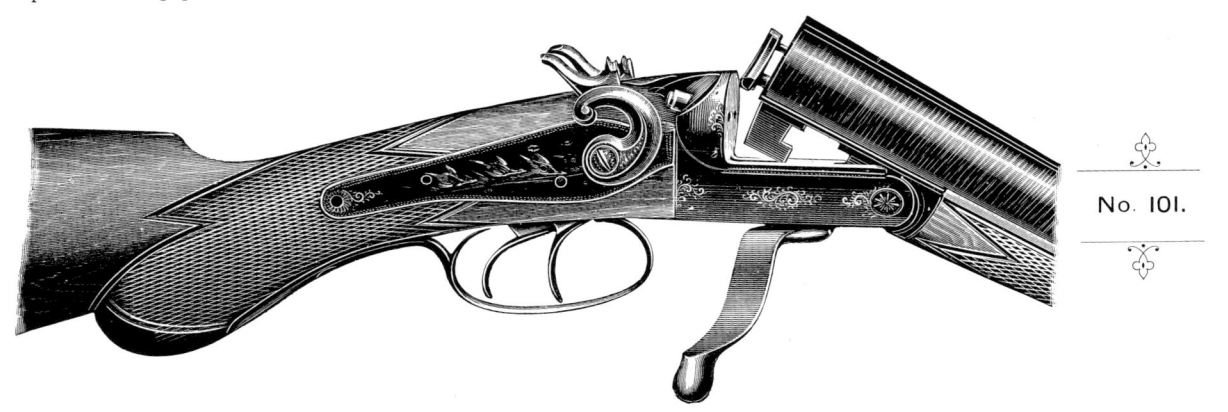

No. 101.

Small Rebounding Hammers below line of sight.

No. 101.—Back Action Locks, Under Lever with Double Grip, English Twist Barrels, Recess Choke, Spring Fore End Fastener, Horn Heel Plate, Pistol Grip Stock.

Nett Cash 50/-

No. 102.

Small Rebounding Hammers below line of sight.

No. 102.—Top Lever Back Action Locks, English Twist Barrels, Recess Choke, Spring Fore End Fastener, Pistol Grip Stock, Horn Heel Plate.

Nett Cash 50/-

Beware of Foreign Rubbish advertised at above prices.

Any Goods not approved of I shall be pleased to exchange if returned in good order three days from receipt of same.

W. J. GEORGE, GUNSMITH, DOVER.

THE ECLIPSE FARMER'S GUN.

This Gun is Designed to meet the wishes of those who require a Well Finished Gun at a Reasonable Price.

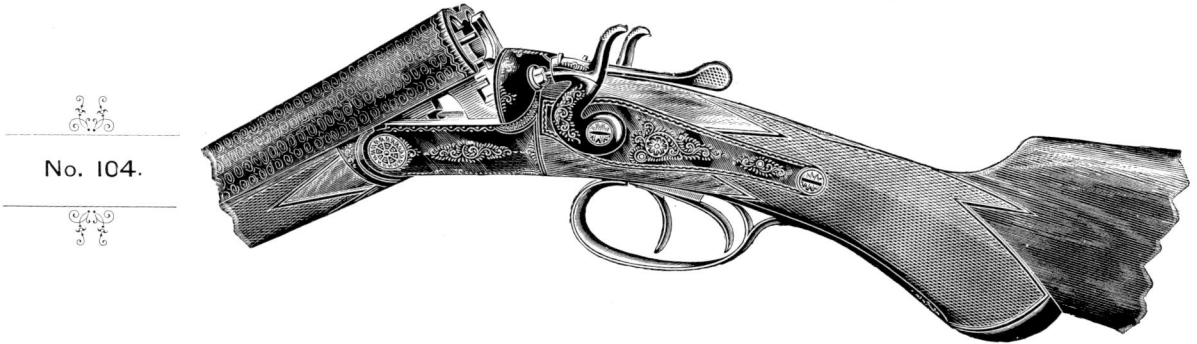

No. 104.

Small Rebounding Hammers below line of sight.

No. 103.—Top Lever, Double Bolted Action, Back Action Locks, with Fine English Twist Barrels, Recess Choke, Best Selected Walnut Pistol Grip Stock, Spring Fore End Fastener, and Horn Heel Plate, £3 : 3 : 0.

No. 104.—Ditto, Ditto, with Damascus Barrels, and Well Engraved, £4 : 4 : 0.

P.S.—Either of these Guns supplied with Lever under Guard, Special Eccentric Action, to Customer's order at same price.

The Eclipse Keeper's Gun.

These Guns are fitted with great care and special attention paid to all wearing parts, so that I have every confidence in saying that they will stand an unlimited amount of hardwear, and special attention is paid to the boring so as to give the best possible results in the shooting powers of same.

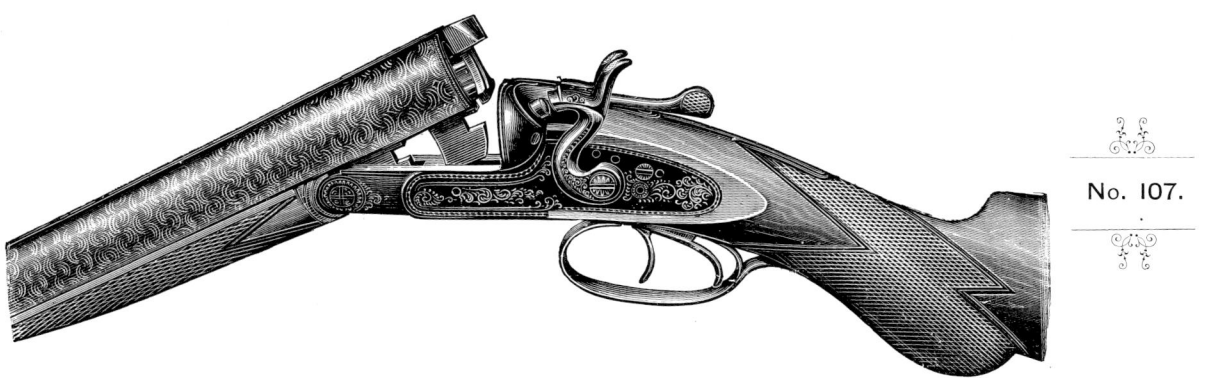

No. 107.

Small Rebounding Hammers below line of sight.

No. 105.—Top Lever, Bar Locks, Double Bolted Action, Twist English Barrels, Pistol Grip Stock, selected Walnut, Spring Fore End, £3 : 5 : 0.

No. 106.—Ditto, with extended Doll's Head Steel Rib, £3 : 12 : 6.

No. 107.—Ditto, Ditto, with Fine Curly Damascus Barrels, £4 : 5 : 0.

Any Goods not approved of I shall be pleased to exchange if returned in good order three days from receipt of same.

W. J. GEORGE, GUNSMITH, DOVER.

THE ECLIPSE COUNTY GUN.
FAR KILLING POWERS GUARANTEED.

This Gun is designed for those who require the best possible Far Killing Game Gun to be had, with all the latest improvements, it is fitted with Stirling Steel Joint Pin Greener's Wedge Steel Cross Bolt, and best possible attention has been paid to fitting all parts to prevent any part becoming loose, it is especially recommended for Wild Fowl Shooting with No. 4. Shot, or for use abroad.

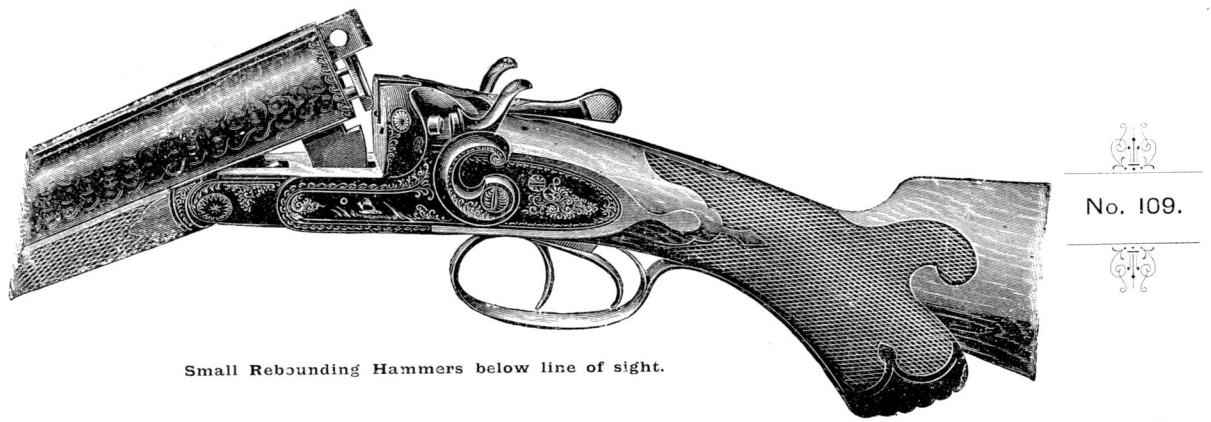

No. 109.

Small Rebounding Hammers below line of sight.

No. 108.—Top Lever, Greener's Treble Cross Bolt Bar Locks, Fine Twist Barrels, Choke Bore, Circular or Round Hammers below line of sight, Pistol Grip Stock, Selected Walnut, Deeley and Edge Fore End Fastener, Horn Heel Plate, £4 : 4 : 0.

No. 109.—Ditto, Damascus Barrels £5 : 0 : 0.

No. 110.—Ditto, Very Fine Damascus or Special Steel Barrels, Superb Finish, to order, £7 : 0 : 0

THE ECLIPSE FEATHERWEIGHT GUN.

This Gun is specially constructed for Gentlemen who like a very light weapon, and is especially suitable for Partridge Shooting at the commencement of the Season, as it only weighs from 5½lbs. to 6lbs.

No. 111.

Small Rebounding Hammers below line of sight.

No. 111.—Top Lever, Greener's Treble Bolted Action, Back Action Locks, Percussion Fence, Left Full and Right Modified Choke, Fine Steel Barrels, Selected Walnut Pistol Grip Stock, or to order, Spring Fore End, £7 : 7 : 0.

No. 112.—Ditto, Ditto, with Extra Finish, Finely Engraved, and Very Best Steel, or Very Fine Damascus Barrels, £10 : 10 : 0.

The above not to be equalled at double the price by Jobbers and Dealers, who advertise Garret-made Rubbish.

Any Goods not approved of I shall be pleased to exchange if returned in good order three days from receipt of same.

W. J. GEORGE, GUNSMITH, DOVER.

THE ECLIPSE PIGEON GUN.

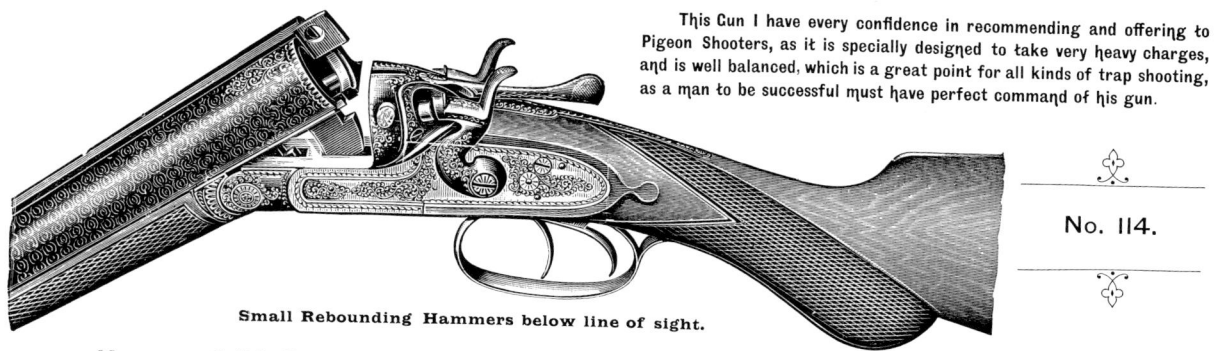

This Gun I have every confidence in recommending and offering to Pigeon Shooters, as it is specially designed to take very heavy charges, and is well balanced, which is a great point for all kinds of trap shooting, as a man to be successful must have perfect command of his gun.

No. 114.

Small Rebounding Hammers below line of sight.

No. 113.—Solid Bar Body, Top Lever, Fine Damascus Barrels, Left Full Choke, or to order, Greener's Treble Cross Bolt Action and Special Joint Pin, all Steel, Well Finished, with Selected Walnut Stock, Pistol Hand, or to order, £6 : 10 : 0.

No. 114.—Ditto, Fine Damascus, or Steel Barrels, Well Finished, and Engraved, £8 : 10 : 0.

No. 115.—Ditto, Special Finish, Pattern Guaranteed at 40 yards, Special High Flat Dead Level File Cut Rib, £10 : 10 : 0.

Under Lever Action if preferred, with Doll's Head Rib Extension, £1 less in either above Models.

The following are selected from a number of unsolicited Testimonials received:—

Mr. W. J. GEORGE,
    Dear Sir,

MINERAL WATER MANUFACTORY,
THE PIER, DOVER,
February, 1900.

The Gun (Special Pigeon) that I bought from you in February, 1898, I must tell you has given entire satisfaction. I have won over £120 in prizes with same at Pigeon Shooting, and all off from 28 to 31 yards rise, which alone speak for its far killing powers and good shooting, and I cannot wish for a better gun.

Yours faithfully,
P. WRAITH.

Mr. GEORGE,
    Dear Sir,

GUILDHALL VAULTS,
GUILDHALL STREET, FOLKESTONE,
October 15th, 1899.

Enclosed please find cheque for Hammerless Gun; I am well satisfied.

Yours truly,
J. TUNBRIDGE.

Mr. GEORGE,
    Sir,

BEACONSFIELD ROAD, DOVER,
January, 1900.

The Gun I bought some time ago at your establishment has turned out a marvellous killer. I have made some most extraordinary long shots, several times I have killed hares which my friends would not shoot at as they thought them too far off.

Yours, etc.,
J. PARSONS, Builder.

Mr. W. J. GEORGE,
    Dear Sir,

THE BURLINGTON,
CASTLE STREET, DOVER,
January, 1900.

The Special Pigeon Gun I purchased from you some time since has given every satisfaction. I have won and shared numerous sweep stakes, and also at Ashford a few weeks since won a pony and trap, value £35, from 26 yards rise, best blue rocks, with 11 consecutive kills from a field of 20 competitors.

Yours truly,
W. SPRATT.

Any Goods not approved of I shall be pleased to exchange if returned in good order three days from receipt of same.

W. J. GEORGE, GUNSMITH, DOVER.

ECLIPSE WILD FOWL AND PUNT GUNS.

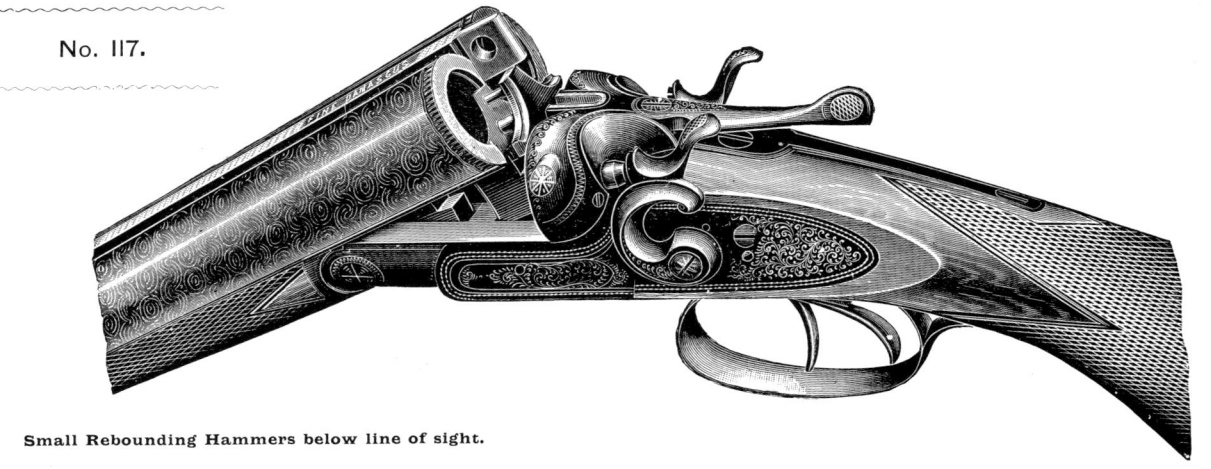

No. 117.

Small Rebounding Hammers below line of sight.

As I make a Speciality of Far Killing Guns, I defy Competitors to equal the Guns described below, either in Far Killing Powers or Value for Money.

No. 116.—English Twist Barrels, Bar, or Back Action Locks, Walnut Stock, Pistol or Straight Hand, Eccentric under Lever Action, made of Stirling Steel, which with Steel Joint Pin effectually prevents the barrels from working loose, Bored for 3in. to 3¼in. Cartridge, 8 Bore 34in. and 36in. barrels, £6 : 6 : 0., 10 Bore 32in. barrels, £4 : 15 : 0.

No. 117.—Ditto, Damascus Barrels, 8 Bore £8 : 0 : 0., 10 Bore, £6 : 0 : 0.

No. 118.—Ditto, but with Top Lever, and Greener's Treble Cross Bolt, 8 Bore, £9 : 0 : 0., 10 Bore, £6 : 10 : 0.

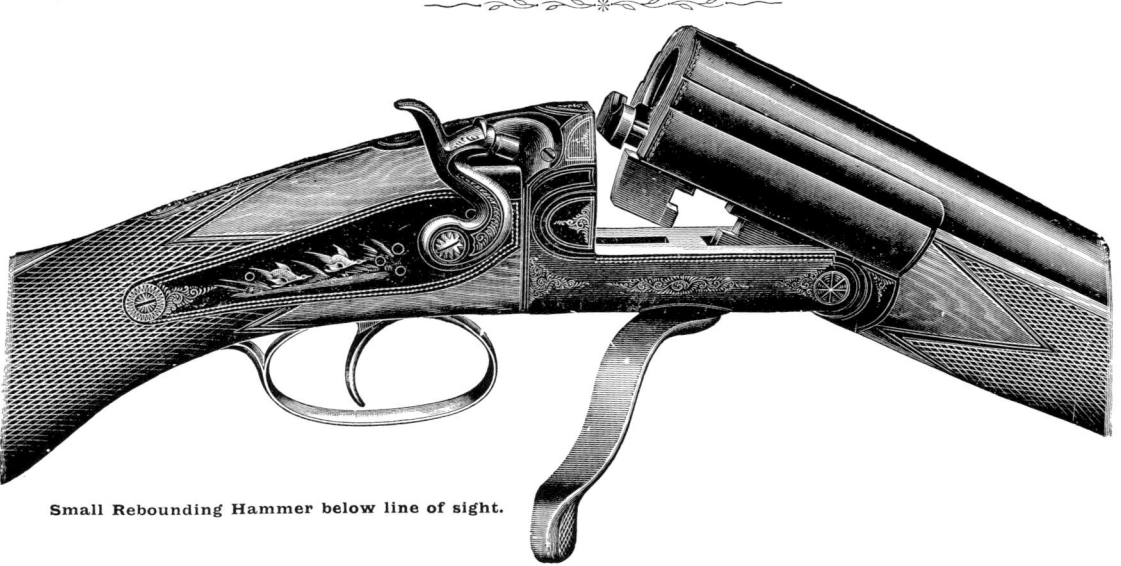

Small Rebounding Hammer below line of sight.

Single Barrel Guns at £1 less than Double, Longer Barrels, to order at 3/- to 4/6 per inch.

I take the liberty of informing Purchasers of the above Guns, that these are all built to Customer's orders. I do not thrust something on you regardless of whether it suits or not, but wish to please my Customers by selling them what will suit them.

Any Goods not approved of I shall be pleased to exchange if returned in good order three days from receipt of same.

W. J. GEORGE, GUNSMITH, DOVER.

THE ECLIPSE HAMMERLESS GUNS.

These Guns I confidently recommend to those who require a Gun to use, and not to look at, the price has been kept as low as possible, without doing away with the very best workmanship and material. I am aware that there are plenty of Foreign and Garret-made Guns advertised at a lower figure, but is it policy to buy at a few shillings less and pay pounds to keep the same in working order?

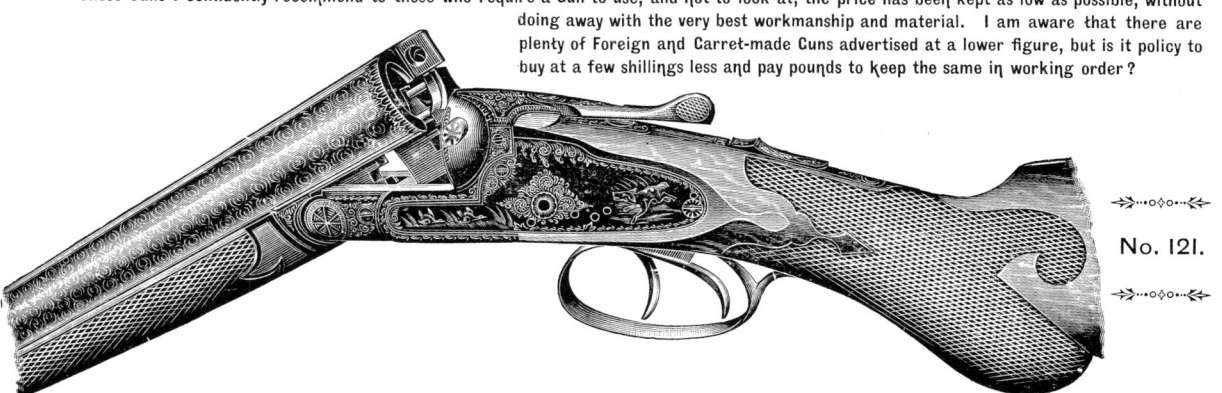

No. 121.

No. 119.—Top Lever, Side Lock, Hammerless, Double Bolt Action, with Stirling Steel Extended Doll's Head Rib, Very Fine Twist Barrels, Choke Bore, Automatic Safety Bolt, Pistol Grip, Selected Walnut Stock, £7 : 7 : 0.

No. 120.—Ditto, Damascus Barrels, £8 : 8 : 0.

No. 121.—Ditto, but with Greener's Treble Cross Bolt, Superior Finish, and Locks Nicely Engraved, £9 : 9 : 0.

No. 124.

No. 122.—Top Lever, Anson and Deeley's System Action, Damascus Barrels, Choked Bored, or order, Greener's Treble Cross Bolt, Selected Walnut Stock, Pistol Hand, and Fitted with my Automatic Safety Bolt, Plain Finish, £10 : 10 : 0.

No. 123.—Ditto, Fine Damascus, or Steel Barrels, Shooting Carefully Regulated, £12 : 12 : 0.

No. 124.—Ditto, Ditto, but Extra Finish, and can be had as a Trap Gun, if required, to fire very heavy charge, £15 : 15 : 0.

ECLIPSE EJECTOR GUNS.

To meet the wishes of numerous Customers, I am prepared to supply either No. 122, 123, 124, Hammerless, Fitted with Patent Automatic Ejector, which ejects exploded Case only, at the extra charge of £5 : 5 : 0.

Any Goods not approved of I shall be pleased to exchange if returned in good order three days from receipt of same.

W. J. GEORGE, GUNSMITH, DOVER.

CONTINENTAL MADE GUNS.

At the request of a few of my Customers I am willing to supply the following Guns, 12 Bore, they have been tested, and the Proof Marks stand good in this country, but no further guarantee can be given.

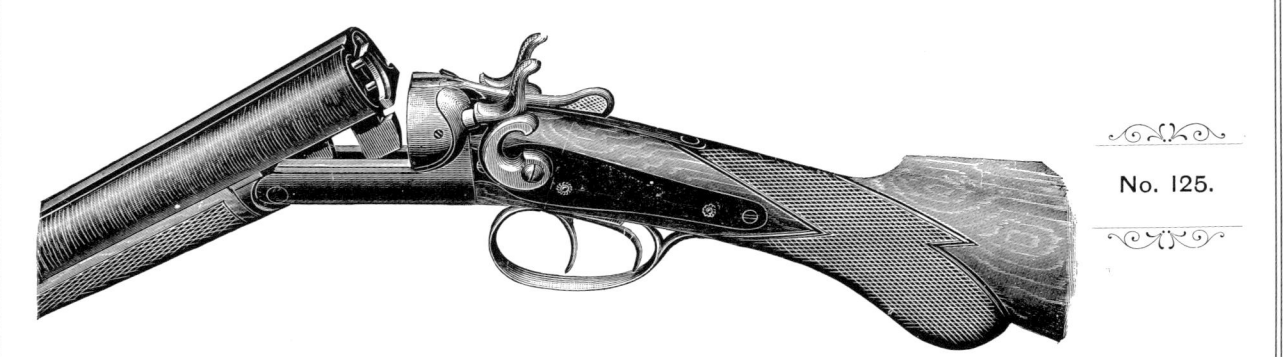

No. 125.

No. 125.—Top Lever Action, Twist Pattern Barrels, Re-Bounding Locks, Back Action, £1 : 17 : 6.

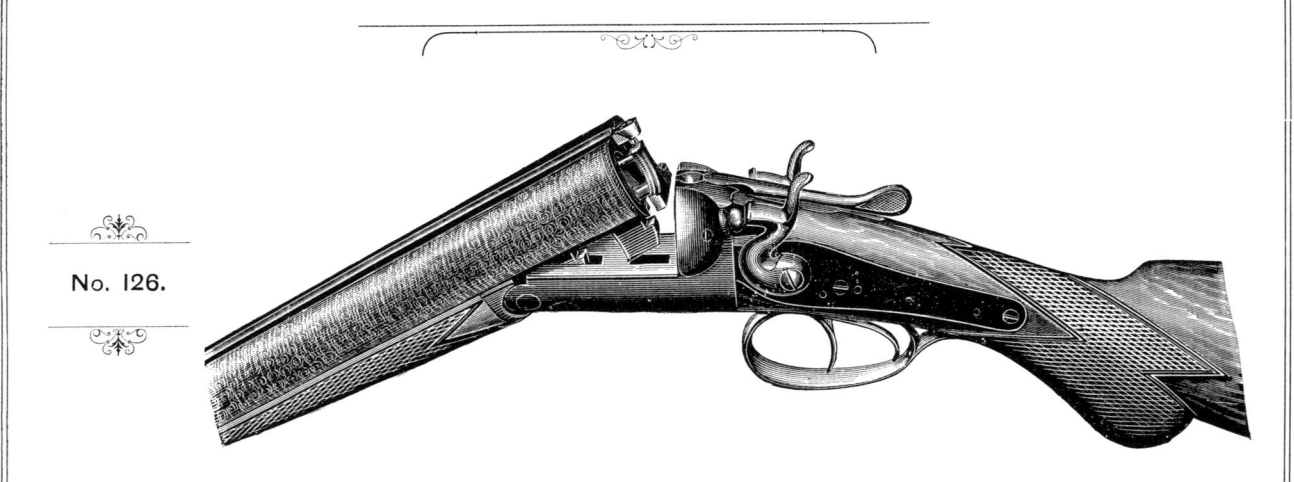

No. 126.

No. 126.—Specification as per No. 1, but Damascus Pattern Barrels, and Extended Doll's Head Rib. £2 : 2 : 6.

No. 127.—Single Barrels as per No. 1, £1 : 16 : 0.

Any Goods not approved of I shall be pleased to exchange if returned in good order three days from receipt of same.

W. J. GEORGE, GUNSMITH, DOVER.

HAMMERLESS SHOT GUNS.

12 BORE ONLY.

CONVERTED FROM GOVERNMENT RIFLES.

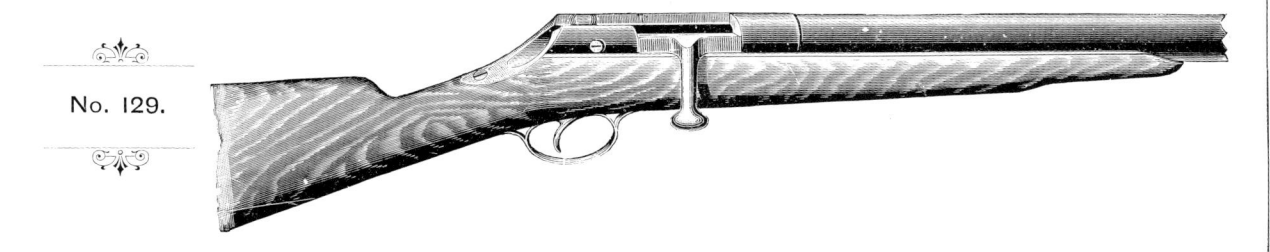

No. 129.

No. 128.—Single Barrel, Steel, Military Browned, Bright Fittings, Walnut Stock, Plain, **17/6.**

No. 129.—Ditto, Better Quality, Checkered, **21/-**

No. 130.—Ditto, Well Finished, Half Pistol Hand, Checkered, **25/-**

The Converted Rifles are all carefully selected, and if at all faulty are thrown out, all the Actions are Tested, and therefore there is not any danger of them going off unawares, as there is with some of this class of Gun which is advertised by Dealers at Cutting Prices, but I do not wish to sell goods I cannot recommend.

Long Range Martini Shot Gun.

12 BORE ONLY.

NOT A CONVERTED RIFLE.

No. 131.—Nickel Plated Action, Half Pistol Hand, Checkered Stock, and Fitted with Automatic Safety Bolt, Guaranteed to make a Splendid Pattern, **42/-**

I supply a Special Cartridge for the above Guns, either Black Powder or Nitro, at 8/- per 100, in 500 Lots, Carriage Paid on Rail to nearest Station.

I can recommend all the above Guns as being made to stand hard wear, and as a Cheap Far Killing Breech Loader, they cannot be surpassed.

Any Goods not approved of I shall be pleased to exchange if returned in good order three days from receipt of same.

W. J. GEORGE, GUNSMITH, DOVER.

THE COLLECTOR'S GUN.
For Ladies, Youths, and Naturalists.

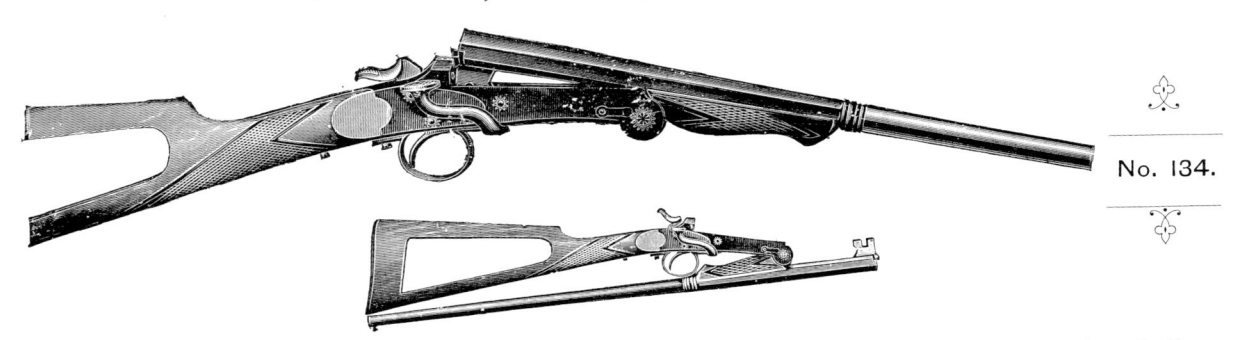

No. 134.

No. 132.—Single Barrel, Blued Steel, Side Lever, Well Made, 410 Bore, £1 : 18 : 0, 28 Bore, £2 : 5 : 0.

No. 133.—Ditto, Better Finished, Pistol Hand Stock, 410 Bore, £2 : 7 : 6, 28 Bore, £2 : 15 : 0.

No. 134.—Single Barrel, Folding, 410 Bore, Gun Side Lever, Fixed Wood Fore End, Blue Steel Barrel, Case Hardened Action, £2 : 10 : 0.

No. 135.—Ditto, Double Barrel, £4 : 0 : 0.

Directions for No. 134:—For ordinary use this Gun opens and shuts like a Side Lever Gun. To fold it up, press in the Button on side of action, and bend barrel forward till in position with barrel touching stock, where it locks itself. To unfold, press the Button again and close the Gun in the usual manner.

THE WANDERER'S GUN.

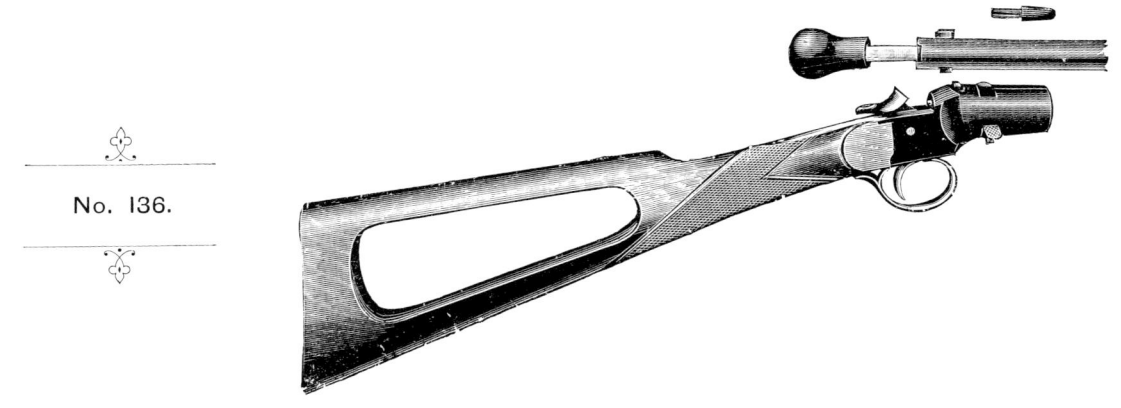

No. 136.

No. 136.—This Gun is specially designed for those who are fond of Shooting, but do not like the idea of always carrying their Gun under their arm, the Barrel therefore is made to fit on the Bayonet principle, the Stock with Lock attached can then be carried in the Coat Pocket, while the Barrel has a handle fitted to the Breech, and a furl in the end for the purpose of using same as a Walking Stick, Selected Walnut Stock, Checkered, Barrel and Fittings Blued Steel, £2 : 2 : 0. Can be had 410, 28, 20 Bore, with Skeleton Stock, as per Block, or 12 Bore, not Skeleton Stock.

Any Goods not approved of I shall be pleased to exchange if returned in good order three days from receipt of same.

W. J. GEORGE, GUNSMITH, DOVER.

ROOK AND RABBIT RIFLES.

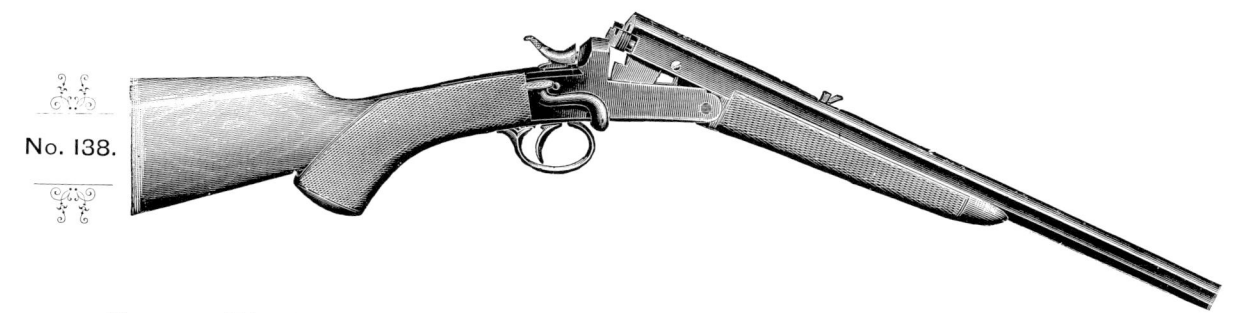

No. 138.

No. 137.—Side Lever, Rifled Barrel, Standard Sights, Octagon Barrel, Well Sighted, Checkered Grip, Blue Action and Barrel, 300 and 380 Bore, £2 : 5 : 0.

No. 138.—Ditto, Very Superior Finish, and Fitted with Ejector, Detachable Fore End, Accurately Sighted, and Shot, £4 : 10 : 0.

No. 139.—Martini Action Rifle, Marbled Fittings, Pistol Hand Stock, Octagon Barrel, Rifled, and Well Sighted, 300 Bore, £3 : 10 : 0.

SALOON AND GARDEN RIFLES.

Accurately Sighted up to 30 Yards Fire, either Shot or Ball No. 1, 2, 3 Cartridges.

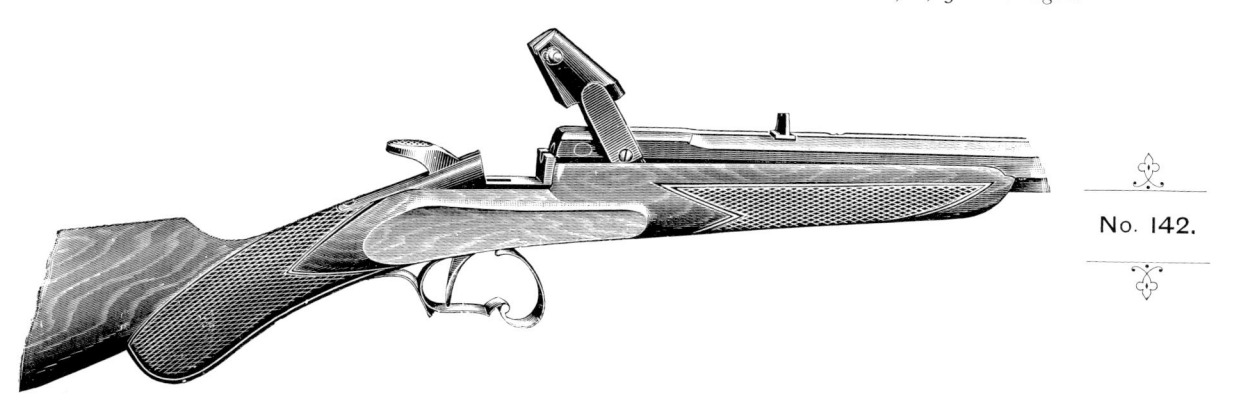

No. 142.

No. 140.—Flobert, with Side Extractor, Standard Sights, Well Sighted, Approved for the No. 1 Shot or Ball Cartridges, 8/6.

No. 141.—Improved Short Block, Fitted with Automatic Extractor, Well Finished, and Nicely Polished Actions, Improved Sighted, Blued Barrel, Fires No. 1, 2, 3 Shot or Ball Cartridges, 13/6.

No. 142.—Special Extra Long Block, Action Ditto, Extra Finish, Pistol Hand, Accurately Sighted, 16/6.

Any Goods not approved of I shall be pleased to exchange if returned in good order three days from receipt of same.

W. J. GEORGE, GUNSMITH, DOVER.

SPECIAL SAFETY CARTRIDGE RIFLE.

No. 143.

FIRES FOUR KINDS OF AMMUNITION.

As noiseless as an ordinary Air Gun, effective up to 200 Yards range, will put a Ball through 4in. of Timber at 100 Yards. It cannot be beaten for Rook and Rabbit Shooting or Target Practice, being most accurate. It can be taken to pieces and put in the pocket in three seconds. Whole length when together 33in., weight 4½lbs., Bore 220.

No. 143.—Plated Barrel and Fittings, Breech Block Hardened in Colours, Splendidly Polished, Walnut Stock, the Barrel is made of Steel and Well Rifled, Automatic Extractor, and Adjustable Rear Sight. Price packed in Case, complete, with Cleaning Rod, £1 : 10 : 0.

No. 144.—The same Rifle, but Fitted with Plated Steel Skeleton Stock, £1 : 6 : 0.

AMMUNITION FOR THE ABOVE.

| No. 1 Ball ... ... 1/- per 100 | No. 22 Cal. (Long) ... 2/6 per 100 |
| ,, 22 Cal. (Short) ... 2/- ,, ,, | ,, 22 Cal. (Shot) ... 3/6 ,, ,, |

Carriage Paid on 500 Cartridges.

India Rubber Heel Plates for this Rifle can be supplied at 2/6 each.

BEWARE OF IMMITATIONS.

WALKING STICK GUNS.

BREECH LOADING.   LONDON PROVED.

No. 146.

These Walking Stick Guns are London Proved, and Guaranteed to be Well Made, and will last out a dozen Foreign made articles which are so often advertised at fancy prices.

No. 145.—320 Bore, Central Fire Gun Metal Barrel, Malacca Cane Covered, Black Horn Handle, Self Cocking, and Fitted with Ejector Action, and an Improved Trigger, Specially Bored to ensure Good Shooting, £1 : 0 : 0.

No. 146.—Ditto, Improved Steel Breech, and Black Horn Handle, £1 : 3 : 6.

No. 147.—380 Bore, all Steel Barrel, Ditto, £1 : 7 : 0.

No. 148.—410 Bore, Ditto, Ditto, £1 : 16 : 0.

Shot Cartridges for the above in Brass Perfect Cases, per 100 : 320 Bore 4/6, 380 Bore 5/6, 410 Bore 6/6. Carriage paid on 500 Cartridges.

Any Goods not approved of I shall be pleased to exchange if returned in good order three days from receipt of same.

W. J. GEORGE, GUNSMITH, DOVER.

BEST ENGLISH AIR GUNS AND WALKING STICK GUNS.
FITTED IN CLOTH COVER WITH INSTRUCTIONS.

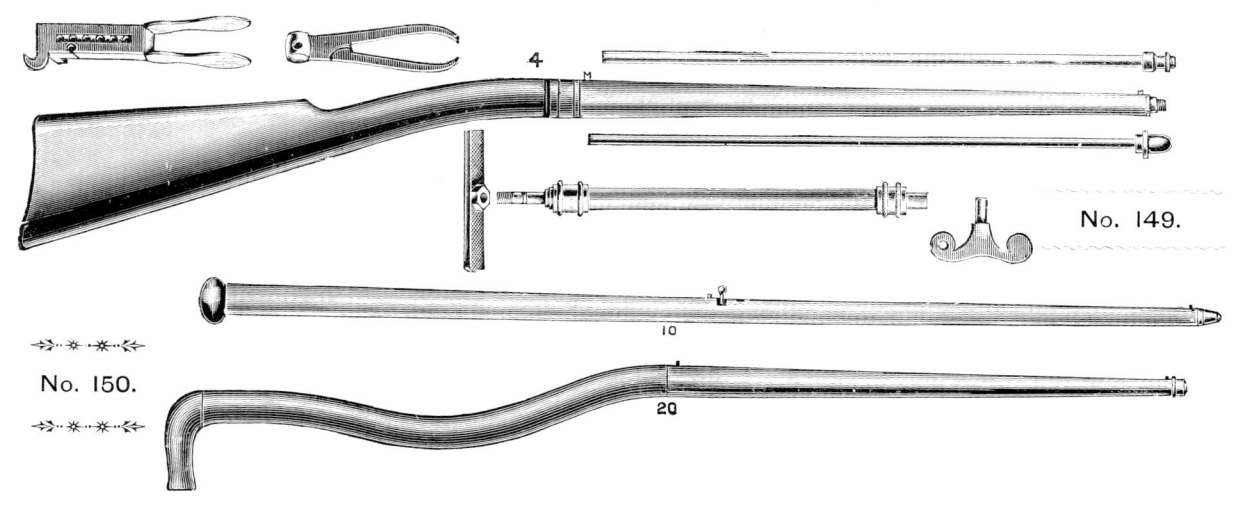

No. 149.

No. 150.

No. 149.—Air Butt Gun, Fitted with Rifle and Shot Barrels, Strong Pump, etc., will Fire Twelve Times with One Charging of Air, complete with Wad Cutter, Bullet Mould, and Shot Measure, Ram Rod, etc., £5 : 17 : 6.

No. 150.—Ditto, Air Cane, £4 : 7 : 6.

No. 151.—Ditto, but Fitted only with Rifle, or Shot Barrel, £3 : 5 : 0.

GENUINE GEM AIR GUNS.
OFTEN ADVERTISED, BUT SELDOM SEEN.   BEWARE OF WORTHLESS IMMITATIONS.

No. 152.

No. 152.—No. 1. Bore, with Plated Actions and Blued Barrel, Walnut Stock, Well Polished, and Highly Finished in every way, Actions all Steel, and Well Fitted, *will last out a dozen Imitations which are so largely advertised*, Accurately Sighted to Fire, Darts for Indoor Practice, and Slugs, or Small Charge of Shot for Small Small Birds, Rats, etc., complete, with Slugs, Darts, and Pinchers, 17/6.

No. 153.—Ditto, but Extra Large and Strong, 27/6.

No. 154.—No. 3 Bore as per No. 152, 19/6.

No. 155.—Combination Gem Air Gun and Rifle, fires No. 2 Slugs, and No. 1 Bullet, or Shot Cartridge, at option, can be Converted from an Air Gun to a Rifle in three Seconds, 38/-

No. 156.—Ditto, but Very Superior, and Fitted with Two Barrels, one inside the other, on the Morris Tube Principle, fires No. 1 Slugs and Darts, and No. 3 Bullet or Shot Cartridges, £2 : 2 : 0.

THE 20TH CENTURY DAISY AIR GUN.
With Detachable Barrel, to fire No. 1 Slugs, or Darts, or B.B. Shot, 5/6.

THE NEW KING AIR GUN.
Fires No. B.B. Shot, or No. 1. Darts. Highly Nickel Plated Barrel, and Actions, with Well Polished Stock, a Splendid Gun for Boys, 5/- B.B. Shot per lb.

No. 1. Slugs 1/- ℔ 1000. No. 3. Slugs 2/2 ℔ 1000. No. 1. Darts 1/3 ℔ doz. No. 3. Darts 2/3 ℔ doz.

Any Goods not approved of I shall be pleased to exchange if returned in good order three days from receipt of same.

W. J. GEORGE, GUNSMITH, DOVER.

Muzzle Loading Guns.

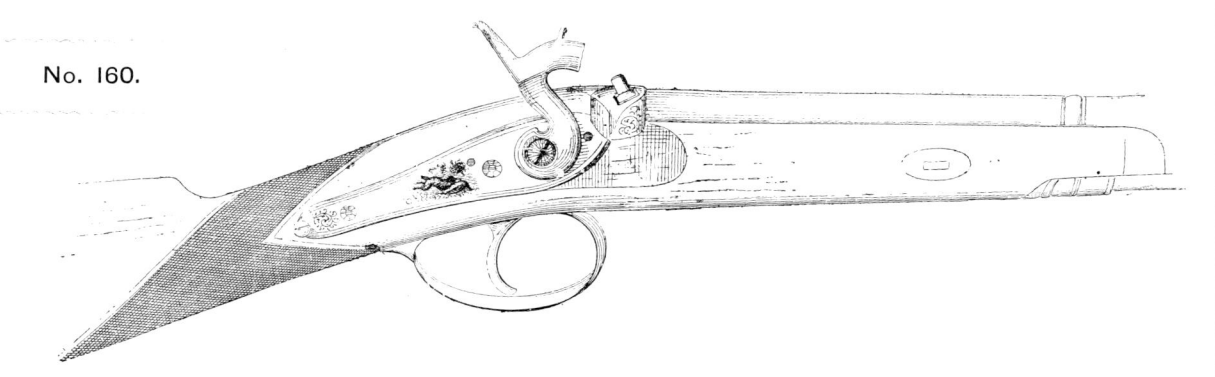

No. 160.

No. 157.—Single Muzzle Loading Gun, Converted from Government Rifles, about 14 Guage, Smooth Bored, Fitted with Steel Rod, **13/6.**

No. 158.—Sham Twist Barrel, Silver Nose Cap, Hard Wood Rod, Full Mounted, Bolted and Escutcheoned, Blued and Hardened Furniture, **16/-**

No. 159.—Single Real Twist Barrel, Patent Breech Break-off, Good Walnut Stock, Well Buffed, Hard Wood Rod, and Well Finished, **£1 : 1 : 0.**

No. 160.—Single, Very Superior Ditto with Shoulders, **£1 : 10 : 0.**

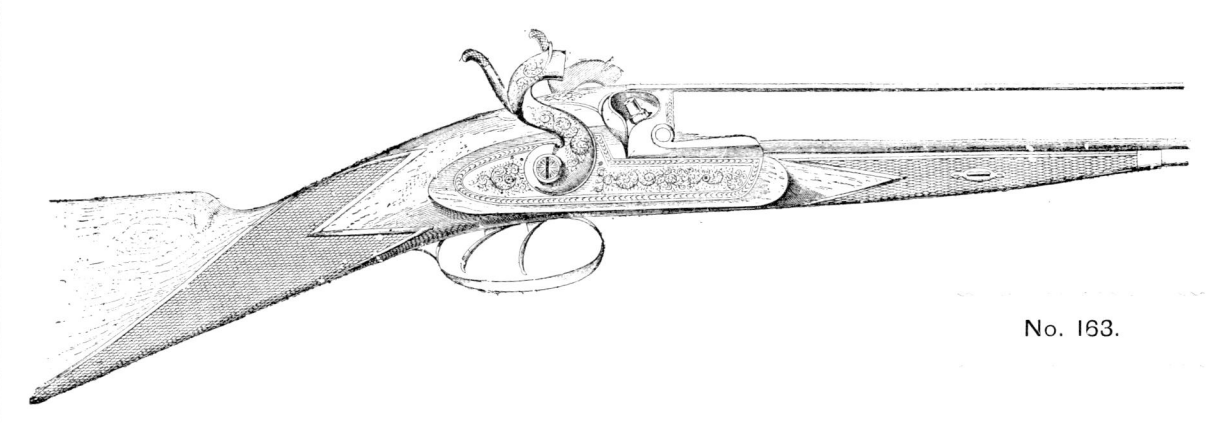

No. 163.

No. 161.—Double Sham Twist, with Patent Breech, Walnut Stock, Well Buffed, Bolted and Escutcheoned, Hard Wood Rod, Brass Tip and Worm, **26/-**

No. 162.—Ditto, but Real Twist Barrels, **35/-**

No. 163.—Ditto, Very Superior, **£2 : 10 : 0.**

Any Goods not approved of I shall be pleased to exchange if returned in good order three days from receipt of same.

W. J. GEORGE, GUNSMITH, DOVER.

Genuine Improved Smith and Wesson Pattern Revolvers.

No. 164.

No. 164.—Self Cocking, and Automatic Cartridge Ejector, Rubber Handle, Splendidly Nickel Plated, or Blued.

320 Bore, **20/-**, 380 Bore, **21/-**, 450 Bore, **22/-**

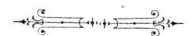

No. 165.

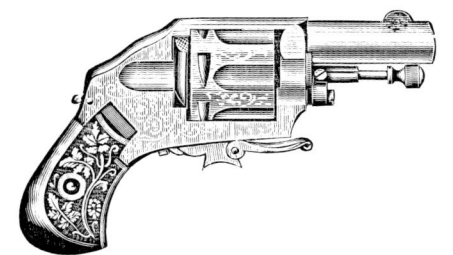

HAMMERLESS REVOLVER.

No. 165.—Hammerless 320 Bore, with Figured Rubber Stock, Engraved, and Splendidly Nickled, **21/-**

Constabulary or Bulldog Revolvers.

No. 166.

No. 166.—Nickel Plated, Well Engraved, with Rubber Handle, Well Finished.

320 Bore, **10/-**, 380 Bore, **11/6**, 450 Bore, **14/-**

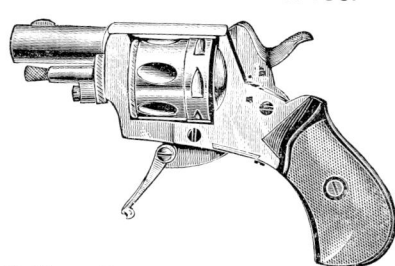

POCKET REVOLVERS.

No. 168.

No. 167.

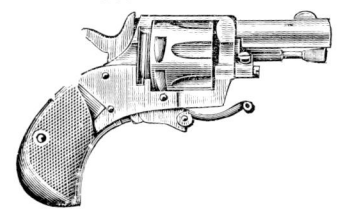

No. 167.—Chequered Walnut Stock, Well Nickeled, or Blued, 220 Bore, **12/6**.

Plush Lined Case for same, **2/6**.

No. 168.—The Lady Pocket Revolver, 220 Bore, Pearl Stock, Gilt Fittings, Marble Hardened, and Plush Lined Case, all Splendidly Finished, Suitable for Presentation, **32/6**.

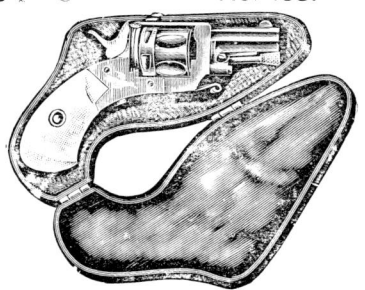

PIN FIRE REVOLVERS.

No. 169.—Ordinary Pattern, Pin Fire, Nickled or Blued, Well Finished, 7M Bore, **5/-**, 9M Bore, **7/-**, 12M Bore, **9/-** Cleaning Rods for Revolvers, **1/-**

Beware of Imitations. Please note the difference between the above Prices and those of my competitors. All the above bear Legal Proof Marks.

REVOLVER CARTRIDGES—BEST.

220, **2/-** ℙ 100. 320, **3/6** ℙ 100. 7M, **3/-** ℙ 100. 380, **4/6** ℙ 100. 9M, **3/6** ℙ 100. 12M, **4/-** ℙ 100. Cordite Powder—476, **7/6** ℙ 100. 455, **7/6** ℙ 100. 450, **6/6** ℙ 100. Black Powder, ditto, **1/-** ℙ 100 less.

Leather Belts, with Holster for Revolver, and Cartridge Pouch attached, **7/6, 10/6, 15/6**.

Any Goods not approved of I shall be pleased to exchange if returned in good order three days from receipt of same.

W. J. GEORGE, GUNSMITH, DOVER.

SALOON PISTOL.

No. 170.

No. 170.—Drop Barrel Saloon Pistol, No. 1. Bore 6in., Blued Barrel, Bright Action, **12/6**.

DERRINGER PISTOLS.

No. 171.

No. 171.—Derringer Pistol, Fires No. 1 or 2 Bullet or Shot Caps, Splendidly Nickeled, and Fitted.

4/- Short Barrel. **5/-** Long Barrel.

No. 172.

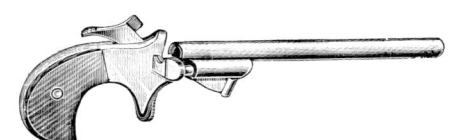

No. 172.—Ditto, Derringer, Fitted with Extractor.

5/6 Short Barrel ; **6/6** Long Barrel.

AMMUNITION FOR THE ABOVE PISTOLS.

BULLETED CARTRIDGES, BEST QUALITY.	SHOT CARTRIDGES, BEST QUALITY.
No. 1. ... 1/- ℔ 100. No. 2. ... 1/6 ℔ 100.	No. 1. ... 1/5 ℔ 100. No. 2. ... 2/- ℔ 100.

No. 1 Noiseless Cartridges, 4/6 for 500.

MUZZLE LOADING PISTOLS.

No. 173.

No. 173.—Single Muzzle Loading Pistol, **2/6** each.

No. 174.—Double Ditto, **4/6** each.

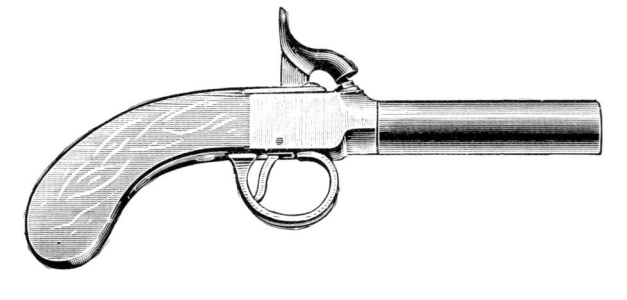

Any Goods not approved of I shall be pleased to exchange if returned in good order three days from receipt of same.

W. J. GEORGE, GUNSMITH, DOVER.

GUN CASES.

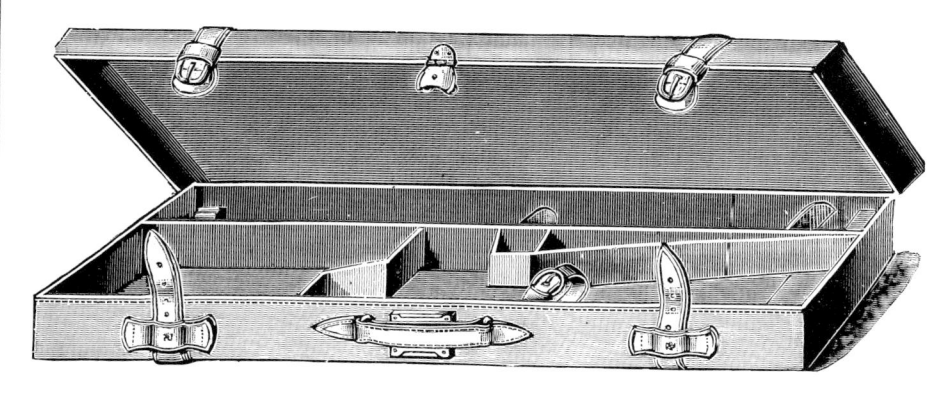

No. 178.

No. 175.—Good Quality, Wooden Case, with Three Divisions, Brass Hooks, Leather Handle, **6/-**
No. 176.—Brown Canvas, Lined Red, with Brass Hooks, and Leather Handle, **10/-**
No. 177.—Ditto, Very Superior, Leather Bound, with Lock and Straps, **14/6.**
No. 178.—Splendid Brown Hide Leather Case, Lined with Red Cloth, with Lock and Straps, **36/-**

LEG OF MUTTON GUN CASES.

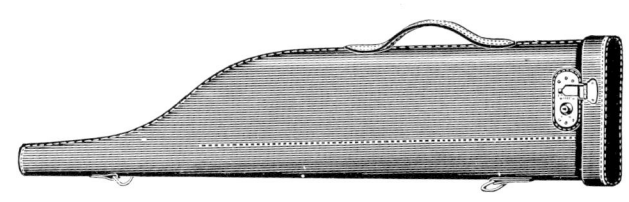

No. 181.

No. 179.—Leg of Mutton Check Strong Waterproof Canvas, **8/6.**
No. 179A.—All Leather Ditto, **14/-**
No. 180.—Ditto, "The Jubilee," Blocked with Handle Loops for Sling, &c., **21/-**
No. 181.—Ditto, Selected, Very Superior Leather, Lined with Mole, and Fitted with Slide Lock, **36/-**
Leather Slings, with Swivel Hooks, **4/-** each.

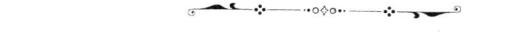

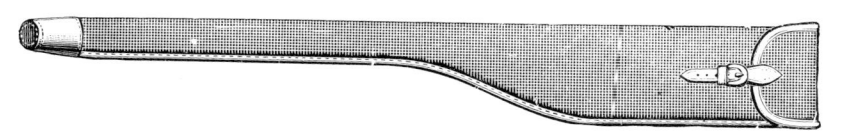

WATERPROOF CASES FOR GUNS.

No. 184.

No. 182.—Light Check, Waterproof, **3/-**
No. 183.—Very Strong, Leather Bound Brown Check, **4/6.**
No. 184.—Ditto, Special Quality, Waterproof Canvas, Leather Bound, and Nose Cap, **6/6.**

Any Goods not approved of I shall be pleased to exchange if returned in good order three days from receipt of same.

W. J. GEORGE, GUNSMITH, DOVER.

GAME BAGS.

No. 185.—Strong Tanned Canvas Bag, Web Slings, with Net, 16in. × 14in. **7/6**.

No. 186.—Ditto, 18in. × 16in., and Flap Fitted, **10/6**.

No. 187.—Ditto, 22in. × 18in., **21/-**

Any of the above Leather Lined, 5/- each, extra.

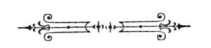

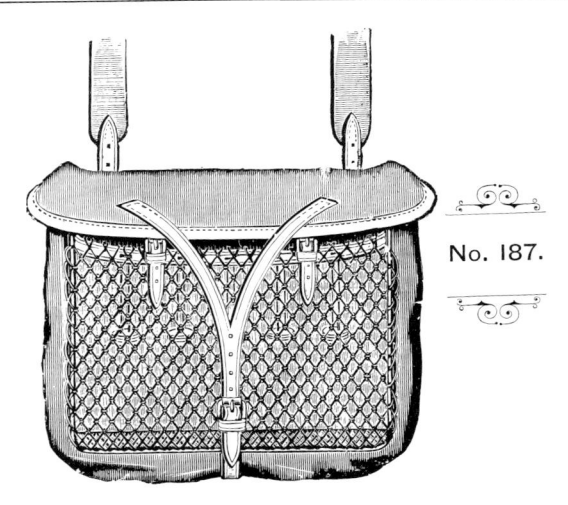

No. 187.

No. 188.

CARTRIDGE BAGS.

	To hold 50.	To hold 75.	To hold 100.
No. 188.—Brown Check Waterproof Canvas	4/-	5/-	6/-
No. 189.— ,, Tan Buck Waterproof, with Pocket	6/-	7/-	8/-
No. 190.—Best Leather throughout	10/6	12/6	14/6

CARTRIDGE BELTS.

	To hold 25.	To hold 30.	To hold 35.
No. 191.—Russet Leather Belt, with Finest Spring Steel Clips	5/-	6/-	7/-

All Leather, Ditto, same prices.

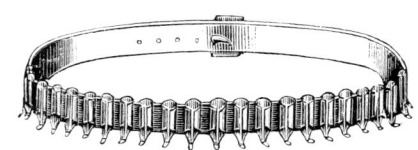

No. 191.

GAME CARRIERS.

No. 192.

No 192.—Polished Oak Carrier, Leather Handle, and Cords, for suspending Birds, **5/-**

No. 193.—Double Carrier, Polished Oak, with Traps, **10/-**

CLEANING RODS.

No. 195.

No. 194.—12 Bore Cleaning Rod, with One Brass Joint, Red Wood, with Wire Brush, Mop, and Brass Jag, **2/6**.

No. 195.—Ditto, Very Superior, with Two Brass Joints, **3/6**.

No. 196.—Steel Rods with Brass Knob, Screwed, and Jag's Fitted, No. 1, **1/6**, No. 2, **1/9**, No. 3, **2/-**

Any Goods not approved of I shall be pleased to exchange if returned in good order three days from receipt of same.

W. J. GEORGE, GUNSMITH, DOVER.

Loading Implements.

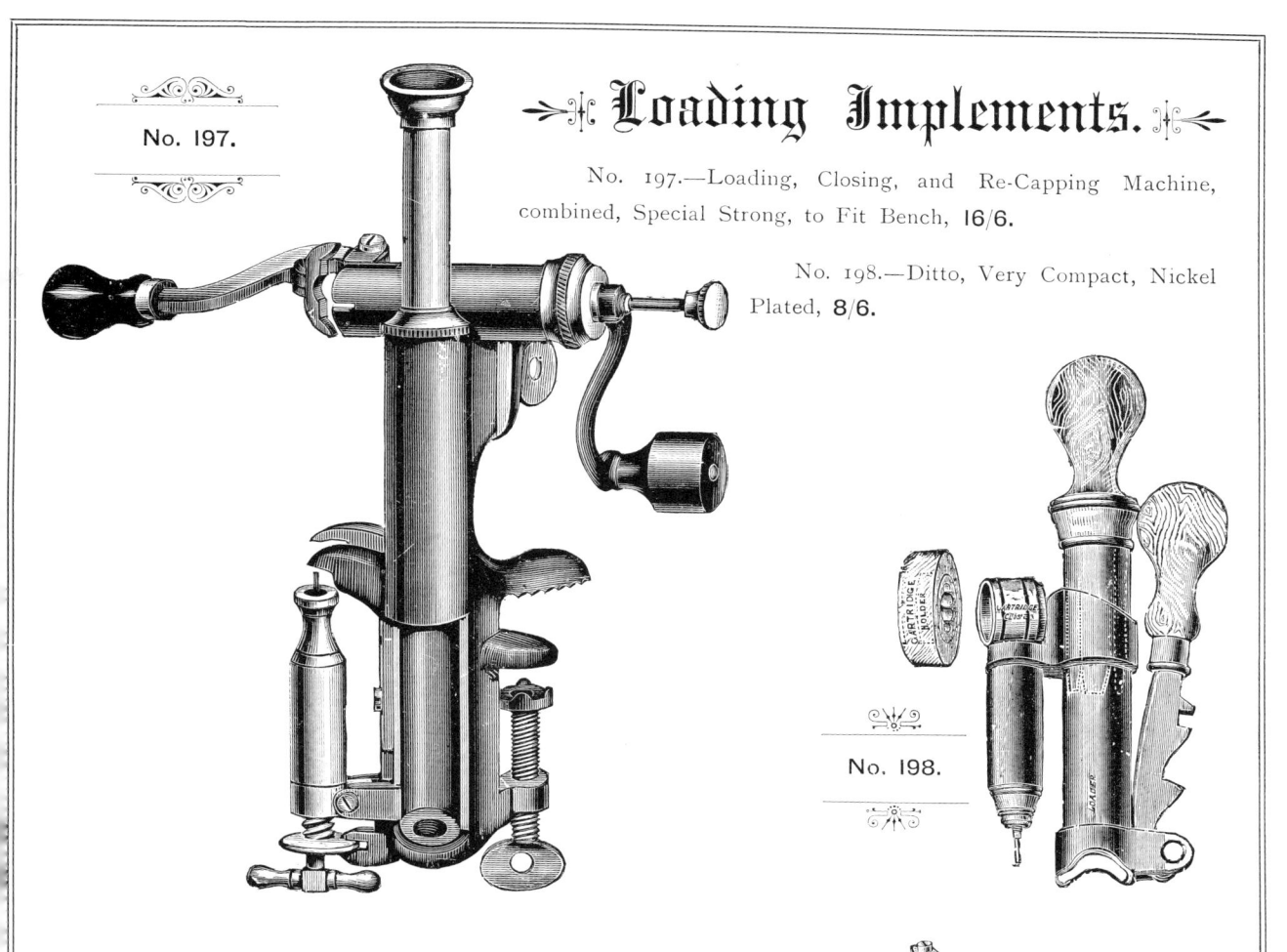

No. 197.

No. 197.—Loading, Closing, and Re-Capping Machine, combined, Special Strong, to Fit Bench, **16/6.**

No. 198.—Ditto, Very Compact, Nickel Plated, **8/6.**

No. 198.

No. 199.—Turnover and Re-Capper combined, Very Strong, to Fit Bench, **9/6.**

No. 200.—Bench Turnover, Very Strong, without Re-Capper, **5/-**

No. 201.—Double Pillar, Brass Re-Capper, **5/-**

No. 202.—Rosewood Re-Capper, Block, and Rammer, **2/9.**

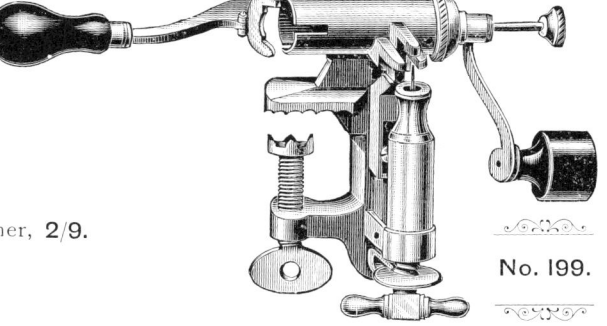

No. 199.

POWDER AND SHOT MEASURES.

No. 203.

No. 203.—Nickel Plated Powder and Shot Measure, **1/6.**

No. 204.—Powder or Shot Measure, with Wooden Handle, **1/3.**

Any Goods not approved of I shall be pleased to exchange if returned in good order three days from receipt of same.

W. J. GEORGE, GUNSMITH, DOVER.

Sporting Gun Accessories, &c.

Bullet Moulds, 1/-   **Wadcutters,** 10 to 20 Bore, 1/-
½ ℔. **Powder Flasks,** 1/6 and 2/6.   2½ ℔. **Shot Pouches,** 1/6 and 2/6.   **Bottle Gun Oil,** 1/-
Gun Barrel Preservers, 3/- per Pair.   Well Finished **Knuckle Dusters,** 1/6 per Pair.
Pocket Barrel Cleaners, in Pouch, Bristle Brush, and Weight, 2/6.
Gun Chamber Cleaner, with Ebony Handle and Brass Cover, 2/6.
Cartridge Extractors, Brass Finger Ring, 1/-
Ditto,   Best Spring Steel, will fit any Bore, 2/-
Extractor with **Dog Whistle,** 2/-   **Gun Turnscrew,** 1/-   **Nipple Key,** 1/-
Improved Catapult, 9d., and 1/-   **Ferret Muzzle,** Improved Spring or Hinge Pattern, 6d. each.
Plain Ferret Muzzles, 3/- doz.   **Ferret Collars,** 3/6 doz.   **Ferret Bells,** 8/- doz.
Mounted Ramrod, Redwood, Tipped and Wormed, 1/3.
Alarm Guns, to fix on trees or posts, 6/-   **Cartridge Case Resizer,** 2/-
Cartridge (according to size) **Magazines,** from 16/-
Best Leather Rifle or Gun Sling Spring Ends, 5/-   **Plain Rifle or Gun Sling Spring Ends,** 2/-
Red Rubber Recoil Pad, to slip over Butt, lengthens Stock, about ¾ in., 5/6.
Pneumatic Face Pad, all Rubber, Slips over Butt, 4/6.   **Life Protectors,** from 1/6 each.

LIVE PIGEON TRAPS,

9in. square, with Open Bar Work Front, Set of Five, made of Best Cold Rolled Steel, for Pigeon Shooting, &c., £4 : 4 : 0.

RABBIT NETS, WIRES, &c.

RABBIT, BIRD, AND VERMIN TRAPS

of every description supplied, small or large quantity.   Prices on Application.

Martini Henry Rifles, Mannlicher Rifles,
Lee Enfield or Lee Metford Rifles, Colts Rifles and Revolvers,
Smith and Wesson Revolvers, Marlin Rifles, Winchester Rifles.

Any of above Rifles I shall be pleased to quote for upon receipt of application, and guarantee prices will be below London Store Prices of same, and all articles guaranteed genuine.

I CAN ALSO SUPPLY COMBINATION RIFLE AND SHOT GUNS FOR COLONIAL USE.

Any Goods not approved of I shall be pleased to exchange if returned in good order three days from receipt of same.

W. J. GEORGE, GUNSMITH, DOVER.

DECOYS, CALLS, & WHISTLES.

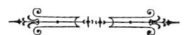

Hawk Kite, Simple, Strong, and Effective, Silk, weight 3ozs., size 2ft. 9in. × 2ft. 6in., Best Material, Paragon Frame, Brass Mounts and Joints, complete in Walking Stick Case	25/-
Ditto, but Calico, Plain Finish, in Bag	17/-
Wooden Decoy, Pigeon, Artistically Painted	5/-
Pneumatic ,, Duck or Drake, ,, ,,	7/-
Wood Pigeon Call	1/9
Duck Call	2/6
Partridge Call	1/-
Grouse Call	1/-
Jay Call	1/-
Plover Call	1/4
Snipe Call	1/8
Cuckoo Call	1/9
Hare Call	1/4
Magpie Call	1/-
Police Whistle	1/8
Best Gamekeeper's Whistle	2/-
Dog Whistle	1/4
Bicycle Whistle	1/3

DOG SLIPS AND COUPLES

OF EVERY DESCRIPTION QUOTED FOR.   PLEASE ENQUIRE FOR WHAT YOU WANT ON POST CARD.

Single Dog Slip, 1/6; Ditto, Improved, 4/-

Patent Single Dog Startes, 8/-; Ditto Double, with Collars, 27/6.

DOG CHAINS, COLLARS, & WHIPS, of every description supplied.

IRON TARGETS.

Figure or Bell, Large Size to hang up, Thick Iron Plate, 8in. wide, 5/-

Ditto, Small Size, 6in., 3/6.

White Cardboard, with Black Bull's Eye, 7/- per 100.

DUMB BELLS, &c.

Sandow's Patent Spring Grip Dumb Bells and Athleitc Specialities supplied.

PRICES ON APPLICATION.

Any Goods not approved of I shall be pleased to exchange if returned in good order three days from receipt of same.

W. J. GEORGE, GUNSMITH, DOVER.

AMMUNITION.

My Sporting Cartridges are all loaded with the best materials by experienced men. Special arrangements being adopted for the proper loading of the Nitro Compound Cartridges.

My whole aim and object being to produce Cartridges that are in the highest degree, effective in pattern, with the least possible recoil.

CARTRIDGES—KYNOCH'S CASES AND LOADED TO ORDER.

Grouse, 13/- per 100; Brown, 7/6 per 100; Blue, 8/- per 100; Brass Perfects, 13/6 100; Red, 10/6 per 100; Red Case, loaded any Nitro, Special Cartridge, 8/- per 100; 16 or 20-guage, 6d. per 100 less; 10-guage, 1/- per 100 more; Long Cases and Heavier Charges, for Trap Shooting, 1/- per 100 extra. Delivered Carriage Paid in Lots of 500.

If a cheap cartridge is required I can supply them at 6/- per 100, in foreign cases, full charge, good wads, and good English coarse grain black powder.

CARTRIDGE CASES—KYNOCH'S. 12 Guage.

Grouse, 6/- per 100; Brown, 3/- per 100; Blue, 3/- per 100; Green, 3/6 per 100; Brass Perfects, 6/6 per 100; Red, for any Nitro, 3/6 per 100. Foreign Cartridge Cases—Green, 2/- per 100; Red, metal lined, 2/6 per 100. Special prices for other sizes.

WADS.—White Felt ... 3/- & 2/6 per lb. in ½-lb. bags.	WADS.—Thin Grease Proof ... 8d. per box.
,, Brown ,, ... 2/- ,, ,,	,, ,, Card ... 8d. ,,
,, Good Hair or Feltine ... 1/6 ,, ,,	,, ,, Overshot, numbered ... 8d. ,,

BLACK GUNPOWDER.

Good for Crow Starving ... 10d. per lb.	Best Treble Strong ... 1/8 per lb.
Best Tower Proof ... 1/5 ,,	Lion, extra strong ... 2/6 ,,
Electric ... 3/6 per lb.	

NEW TRIPLE WADS.—Special Grease Proof Card, White Felt, and Plain Card, all in one wad, for loading purposes, 3/- per lb.

SMOKELESS GUNPOWDER.

Schultze Smokeless ... 5/6 per lb.	S.S. ... 5/6 per lb.
E.C. ... 5/6 ,,	Amborite Smokeless ... 5/6 ,,
Ditto, No. 3 ... 7/6 ,,	Coopal's ,, ... 5/6 ,,

15-lbs. Smokeless or 25-lbs. Black, delivered free; when smaller quantities are ordered the carriage to be paid by the Buyer.

☞ No other goods can be sent in the same package with the powder.

SHOT.—Soft, 3d. per lb.; Chilled, 4d. per lb. Prices for quantities quoted on application.

ROOK RIFLE CARTRIDGES.—BEST.

No. 300 Central Fire, 5/6 per 100; 380 ditto, 5/6 per 100; 297/230 Morris Tube, 3/- per 100.

REVOLVER CARTRIDGES.—BEST.

320, 4/- per 100; 380, 5/- per 100; 450, 5/6 per 100; 7M, 3/- per 100; 9M, 3/6 per 100; 12M, 4/6 per 100; No. 22, U.M.C. American Cartridges, short, 2/- per 100; No. 22, U.M.C. American Cartridges, long, 2/6 per 100; No. 22, U.M.C. American Cartridges, shot, 3/6 per 100; No. 1, U.M.C. Cartridges, 1/- per 100.

WALKING STICK CARTRIDGES.

Best Ball Cartridges, with brass cases ... 320-bore, 4/-; 380-bore, 5/-; 410-bore, 5/6 per box.
,, Shot ,, ,, ,, ,, ... 320-bore, 5/-; 380-bore, 6/-; 410-bore, 7/- ,,

AIR GUN SLUGS AND DARTS.

Best thick steel-pointed darts, No. 1, 1/3 per dozen; No. 3, ditto, 2/3; No. 1, best slug, 1/- per box of 1000; No. 3, ditto, 2/2; Shot, per lb., B.B., 4d.

SPECIAL NON EXPLOSIVE AIR GUN SHOT CARTRIDGES.

No. 1, per 100 box, 1/-; No. 3, ditto, 2/-

SALOON CARTRIDGES.

No. 1. Ball, 1/- per 100; No. 2, Ball, 1/6 per 100; No. 3, Ball, 2/6 per 100; No. 1, Shot, 1/6 per 100; No. 2, Shot, 2/6 per 100; No. 3, Shot, 3/- per 100; No. 1, Noiseless Cartridges, 4/6 per 500.

Good close shooting can only be obtained by using cartridges loaded rationally, and there must be good wadding between the powder and the shot. The secret of good shooting is in the employment of a first-class felt wad over the powder; The texture must be close and firm, but the relative hardness or softness of the wad is of less moment. It should be of the same diameter as the internal diameter of the cartridge case in which it is to be used, and it should be well pressed.

Carriage Paid only on 500 lots of Loaded Cartridges, and 1000 lots of Cartridge Cases.

Any Goods not approved of I shall be pleased to exchange if returned in good order three days from receipt of same.

W. J. GEORGE, GUNSMITH, DOVER.

Instructions for Measuring

Bend and Length of a Stock for Gun or Rifle, when special dimensions are required.

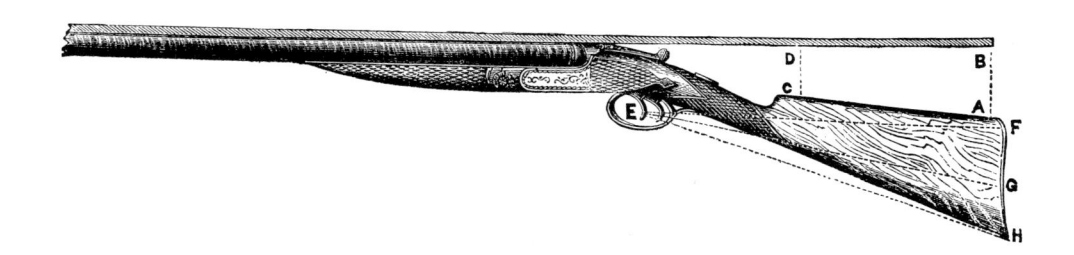

Fasten a straight piece of wood, measuring the full length of the Gun, to the rib of the barrel, so as to extend beyond the stock, and having placed the right hand lock to half cock, take dimensions as follows, viz.:—

Bend, from	**A** to	**B**	Inches.
,,	,,	**C** ,,	**D**	,,
Length	,,	**E** ,,	**F**	,,
,,	,,	**E** ,,	**G**	,,
,,	,,	**E** ,,	**H**	,,

PLEASE DO NOT CUT OUT THIS PAGE WHEN ORDERING, I ONLY REQUIRE THE DIMENSIONS IN YOUR LETTER.

INDEX.

	PAGE.
Accessories, Sporting	20-21
Air Guns—Best English make	13
,, Genuine Gem	13
,, The New King	13
,, 20th Century Daisy	13
Alarm Guns	20
Ammunition	12-13-15-16-22
,, Air Gun	13-22
,, Pistol	16-22
,, Revolver	15-22
Bags, Cartridge	18
,, Game	18
Belts, Cartridge	18
Bells, Ferret	20
Bird, Decoys	21
,, Calls	21
Bulldog Revolvers	15
Cartridges	22
Cartridge Bags	18
,, Belts	18
,, Extractors	20
Carriers, Game	18
Catapults	20
Cleaning Rods	18
Cleaners, Gun Barrel and Chamber	20
Collars, Dog	21
,, Ferret	20
Collector's Gun	10
Constabulary Revolvers	15
Continental made Guns	8
Converted Guns	9
Conversions	INSIDE FRONT COVER
Cycles *(see Special List.)*	
Decoys	21
Dog Chains	21
,, Collars	21
,, Slips and Couples	21
,, Whips	21
,, Whistles	20-21
Dumb Bells	21
English made Air Guns	13
,, ,, Walking Stick Guns	12-13
Excelsior Guns	2
Ferret Bells	20
Game Bags	18
,, Carriers	18
Gun Cases	17
Guns—The Collector's	10
,, Continental made	8
,, Converted	9
,, The Eclipse County	4
,, The Eclipse Ejector	7
,, The Eclipse Farmer's	3
,, The Eclipse Featherweight	4
,, The Eclipse Hammerless	7
,, The Eclipse Keeper's	3
,, The Eclipse Pigeon	5
,, The Eclipse Wild Fowl and Punt	6
,, The Excelsior	2

	PAGE.
Guns—Hammerless Shot	9
,, How to take care of	INSIDE BACK COVER
,, How to Measure	23
,, Muzzle Loading	14
,, Long Range Martini Shot	9
,, Walking Stick	12-13
,, The Wanderer's	10
Hammerless Shot Guns	9
,, Revolvers	15
Hints on Shooting	INSIDE BACK COVER
How to take care of your Gun	,, ,,
How to Measure Rifles and Guns	23
Introduction	1
Implements, Loading	19
Knuckle Dusters	20
Lee Enfield Rifles	20
Lee Metford Rifles	20
Leg of Mutton Gun Cases	17
Loading Implements	19
Mannlicher Rifles	20
Marlin Rifles	20
Martini Shot Guns	9
,, Rifles	20
Measuring Rifles and Guns	23
Muzzle Loading Guns	14
Muzzle Loading Pistols	16
Muzzles, Ferret	20
New King Air Guns	13
Pin Fire Revolvers	15
Pistols—Derringer	16
,, Muzzle Loading	16
,, Saloon	16
Pocket Revolvers	15
Powder and Shot Measures	19
Rabbit Nets, Wires, Traps, &c.	20
Repairs	INSIDE FRONT COVER
Revolvers	15-20
Rifles	20
,, How to Measure	23
,, Rook and Rabbit	11
,, Saloon and Garden	11
,, Special Safety Cartridge	12
Saloon Pistols	16
Saloon and Garden Rifles	11
Shot Guns, Hammerless	9
Shot Measures	19
Slugs and Darts for Air Guns	13-22
Smith & Wesson Revolvers	20
Special Safety Cartridge Rifle	12
Sporting Accessories	20-21
Targets	21
Terms of Business	INSIDE FRONT COVER
Traps—Live Pigeon, Bird, Rabbit, Vermin, &c.	20
Twentieth Century Air Gun	13
Walking Stick Guns	12-13
Wanderer's Gun	10
Waterproof Gun Cases	17
Winchester Rifles	20
Whistles—Bicycle, Gamekeeper's, Dog, Police	21

HOW TO TAKE CARE OF YOUR GUN.

To clean a Gun after a day's shooting. If it be wet, wipe it dry at once, but if the time is too short to do it properly you had better put some oil on the gun, you can leave the cleaning until the next morning, when you can give it your thorough attention. The barrels may be cleaned with the cleaning rod, used with tow and turpentine. To remove the fouling, place the muzzle upon the floor and press down to within an inch of the same and then withdraw it up the chamber, repeating this two or three times and then push it right through. Use the rod with plenty of flannel, and finish with the mop soaked in neatsfoot or Rangoon oil. Always wipe the barrels dry and clean before oiling, and do not use the oiling mop for a foul barrel.

The cocking lifters of hammerless guns, the triggers and holding-down and top bolts, if inclined to clog, may be oiled with a feather dipped in petroleum. To lubricate them use the best neatsfoot. Should the barrels at any time become rusty, either inside or out, scald them well with a kettle of boiling water, afterwards wiping them perfectly dry and oiling them.

Never send a good gun to anyone but an experienced man, as jobbers often place their name on it and tell you they have improved its shooting; and if at any time you want it "choke bored," or converted into a cylinder, send it to no one but the maker or a first-class gunsmith.

Should a gun need repairing always send the whole of it, as to do the work properly, it is necessary that the workman should have it in its entirety, and, if possible, have your gun overhauled by its maker or a good gunsmith every year.

PLEASE NOTE.—I Manufacture the "DOVER" CYCLES,

AND AM ALSO DISTRICT AGENT FOR

HUMBERS, CENTUARS, SINGERS, &c.,

ALSO ALL CYCLE ACCESSORIES.

Prices quoted on application, and Machines sent, Carriage Paid, to nearest Railway Station.

HINTS ON SHOOTING.

Let the learner who desires to become a fair shot first get a suitable gun, and go out at every opportunity, bringing his gun up quickly and firmly to the shoulder to any living and moving object he may come across, taking aim at and following the same but not firing. Let him continue to do this for ten or fifteen minutes, firing a shot now and then when he has the object well covered, but keep the gun on the mark when pulling. Practice of this kind for a week or so will ensure his becoming a shot in a very short time. The sportsman must have confidence and keep cool and watch where his shot strikes, and let him be certain he can kill with the first barrel before using the second.

To him who wishes to improve his shooting I should recommend his going out by himself, taking one dog alone with him to hunt up the birds; and if he has no one to mark his misses, and carries his own game, he will soon become a steady shot, unless his sight is defective, and thus be able to hold his own in a crowd where formerly he felt very nervous.

Taking aim at a bird flying across or at right angles to the sportsman, take aim at the head and keep the gun on the move during the whole time and until after you have pulled the trigger. A knowledge of the velocity of shot enables one to see what allowance is necessary to be made for game moving at a great speed. Take, for instance, a bird coming with the wind and across the line of fire at right angles at a distance of 40 yards. The speed at which the bird is travelling is about 42 miles an hour, or 65 feet per second. Shot travelling at an average rate of 860 feet a second will reach the line of a bird's flight in about 0.15 second, but in that time the bird will have travelled about 8 feet; hence it is plain that if the gun is brought to rest when the trigger is pulled (as is the habit of many) the aim must be taken about 8 feet in front of the bird.

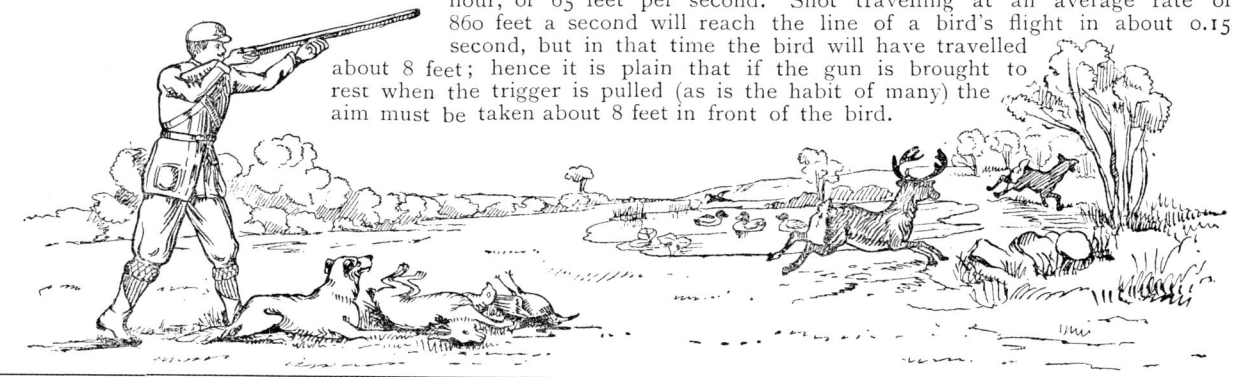

Appendix Two

The Gunmakers of the United Kingdom in 1900

The only comprehensive listing of gunmakers in the British Isles is to be found in the street directories. The over-riding problem of this source is that, beyond the wishes of the promoter of the business, there were no conditions of entry to this list. So it is that the whole spectrum of the trade is represented, from the most prestigious and innovative of firms, down to small town ironmongers who had a few cheap imported guns in stock and sold cartridges of the lowest quality.

It is important to realise the inherent weakness of this list and be aware of the possibility of human error on the part of the compilers. Unfortunately, the closest date to 1900 for the Irish makers was *Slater's Directory* of 1894. For the rest of the United Kingdom, the list derives from *Kelly's Iron Trades Directory* of 1900.

The following role is reproduced exactly as published in these two sources:

Abingdon Works Co. Lim., 94, 95 & 96, Bath St. & Shadwell Street, B'ham.
Adams & Co., 35e, Queen Victoria St., E.C.
Adams & Tait, 1, New Buildings, Price St., B'ham.
Adkin, Henry & Sons, 57, High St., Bedford.
Adsett, Thomas, 101, High St., Guildford.
Agnew & Son, 79, South St., Exeter.
Agnew, Hugh, Newtown Butler, Co. Fermanagh.
Akrill, Henry Esau, Market Place, Beverley, Yorks.
Allan, Arthur, 144, Trongate St., Glasgow.
Allport, John, Bird St., Lichfield, Staffs.
Altendorf & Wright, Vesey St. & Loveday St., B'ham.
Anderson, Henry F., Market Place, Bedale, Yorks.
Anderson, John, 52, Market Place, Malton, Yorks.
Andrews, Charles William Lim., 13, Bath St., B'ham.
Andrews, Charles Wm., 5 & 6, Great Winchester St., E.C.
Andrews, Thomas, 31, New Road, Woolwich, Kent.
Anson, E. & Co., 14, Steelhouse Lane, B'ham.
Arms & Ammunition Manufacturing Co. Limited, 143, Queen Victoria St., E.C.
Armstrong & Co., 115 & 117, Northumberland St., Newcastle-on-Tyne.
Armstrong, Stevens & Son, Whittall St., B'ham.
Atkin, Henry, 2, Jermyn St., S.W.
Atkinson, William & Sons, 11, Skipton St., Morecambe, Lancs.
Atkinson, William, 20, Market St., Lancaster.
Atkinson, William, 58 Highgate, Kendal.
Austin & Son, London Road, Hailsham, Sussex.
Austin, T.C. & Co. Limited., 71, High St., Ashford, Kent.
Baker, Joseph, Norwich St., Fakenham, Norfolk.
Barber, Joseph Henry, 6, Court St., Faversham, Kent.
Barham, Henry, Sun Street, Hitchin, Herts.
Barnes, Fred. & Co., 15, Lionel St. & 57, Livery St., B'ham.
Barnes, Frederick & Co., 109, Fenchurch St., E.C.
Barnett, J.E. & Sons, Duncan St., Leman St., E.C.
Barratt & Son, 48 & 49, High St., Burton-on-Trent.
Bartram, George T., 33 & 35, Bank St., Braintree, Essex.
Bate, George, 132, Steelhouse Lane, B'ham.
Bates, George, Sea Side Road, Eastbourne.
Baxendale & Co., Miller St., Manchester.
Bayley, Thomas, 17½, Whittall St., B'ham.
Beavon, Edward, 1, Court, Whittall St., B'ham.
Beesley, Frederick, 2, St. James' St., S.W.
Bell, Henry, 23, New Buildings, Price St., B'ham.
Bell, Sam'l, 26, Bishop Street, Londonderry.
Benbow, John Griffiths, Bailey St., Oswestry, Salop.
Bentley & Playfair, 315, Summer Lane, B'ham.
Bentley & Playfair, 60, Queen Victoria St., E.C.
Berry, Joseph, Bridge Place, Worksop, Notts.
Bircham, Charles Octavius & Son, 124, Poplar High St., E.
Birmingham Gun & Cycle Co. (The), 15, St. Mary's Row, B'ham.
Birmingham Metal & Munitions Co. Limited (The), Adderley Road, Saltley, B'ham.
Birmingham Small Arms & Metal Co. Limited, 5 & 6, Great Winchester St., E.C.
Blair, John F., 69, Bothwell St., Glasgow.
Blake, James, 14, Square, Kelso, Roxburghshire.
Blakemore, Edward, 8, Sand Street, B'ham.
Blakemore, Thomas, 16, Weaman St., B'ham.
Blakemore, V. & R., 86, Leadenhall St., E.C.
Blanch, John & Son, 29, Gracechurch St., E.C.
Bland, Thomas & Sons, 2, King William St., Strand.
Bland, Thos. & Son, 41, 42 & 43, Weaman St., B'ham.
Blenheim Engineering Co. Limited, (The), Blenheim Works, Eagle Wharf Road, Hoxton, N.
Blissett & Son, 38, South Castle St., Liverpool.
Bond, George Edward, Castle St., Thetford, Norfolk.
Bonehill, Christopher Geo., Belmont Row, B'ham.
Boss & Co., 73, St. James' St., S.W.
Boston, Joseph, 18, Wood St., Wakefield.
Boswell, Charles, 126, Strand, W.C.
Bott, James & Son, 124½, Steelhouse Lane, B'ham.
Bourke, Thos. M., 120, George St., Limerick.
Bourne, Joseph & Son, 7, St. Mary's Row, B'ham.
Bradbury, Thomas, 61, Campo Lane, Sheffield.
Braddell, Joseph & Son, 21, Castle Place, Belfast.
Bradford, R., 16, Quay St., Clonmel, Tipperary.
Braendlin Armoury Co. Limited, 55, Loveday St., B'ham.
Brain, Thomas, 76, Slaney St., B'ham.
Brewster, James, Middleton St., Wymondham, Norfolk.

Brewster, Jas. B., Stratton St. Mary, Long Stratton, Norfolk.
Britcher, Bernard V., 4, Market Buildings, Maidstone.
Burns, George, 31, Loveday St., B'ham.
Burrow, James, 116, Fishergate, Preston.
Burton, Frederick Matthew, 7, Purfleet St., Lynn, Norfolk.
Butt, Edward, Ansty, Salisbury.
Calder, William, 30, Guild St., Aberdeen.
Calderwood & Son, 14, Earl St. North, Dublin.
Carr Brothers, Chancery Lane, Huddersfield.
Carr, E.P., 4, Lowever Parliament St., Nottingham.
Carr, James & Sons, 10 & 11, St. Mary's Row, B'ham.
Cashmore, William, 130, Steelhouse Lane, B'ham.
Chamberlain, Arthur, 18, Queen St., Salisbury.
Chamberlain, Edwin, 1, Bridge St., Andover, Hants.
Chambers, Robert, 39 & 40, Walcot St., Bath.
Chambers, Septimus, 63, Broad St., Bristol & 21, Castle St., Cardiff.
Chapman, John, Foulsham, Dereham, Norfolk.
Chappell, Mrs. E., Thoroughfare, Harleston, Norfolk.
Churchhill, Edwin John, 8, Agar St., Strand, W.C.
Clabrough, J.P. & Johnstone, St. Mary's Row, B'ham.
Clark, William, 73, Bath St., B'ham.
Clarke, Frank, Gothic Arcade, Snow Hill, B'ham.
Clarke, Henry & Sons, 38, Gallowtree Gate, Leicester.
Clarke, Henry W., 12, Queen St., Newton Abbot, Devon.
Climie, Robert, 20, West Blackhall St., Greenock.
Clough, Thomas & Son, 52, High St., Lynn, Norfolk.
Cock, John H. & Co., Market Place, & Cricklade St., Cirencester, Glo'stersh.
Cogswell & Harrison Limited, 226, Strand, W.C. & 141, New Bond St., W.
Cole & Son, 116, Peascod St., Windsor.
Cole & Son, 14, Market Place, Devizes, Wilts.
Collie, William, 10, High St., Montrose, Forfarshire.
Colt Gun & Carriage Co. Limited, 34, Victoria St., S.W.
Colts' Patent Fire Arms Manufacturing Co., 20, Glasshouse St., Regent St., W.
Conyers, Arthur, East St., Blandford, Dorset.
Conyers, John & Son., Market Place, Pocklington, Yorks.
Cooper, Charles H., 67, Cliveland St., B'ham.
Cox & Macpherson, 62, High St., Southampton.
Cox & Son, 28, High St. & 7, Bernard St., Southampton.
Cox, Daniel, 92, Bartholomew St., Newbury, Berks.
Cox, Samuel, 16½, Weaman St., B'ham.
Cranmer, Samuel, Melton, Woodbridge, Suffolk.
Crockart, David & Co., 35, King St., Stirling, N.B.
Crockart, Jas. & Son, 26, Allan St., Blairgowrie, Perthsh.
Cufflyn & Co., 11, Guildhall St., Folkestone.
Cullen, W., Leinster St., Athy, Co. Kildare.
Culling, Thomas, 19, Little Church St., Wisbech, Cambs.
Dadley, Thomas Alex., Market Place, Stowmarket, Suffolk.
Dainteth, Thomas, 121, Bridge St., Warrington.
Darlow, Walter, 27, Midland Road, Bedford.
Davie, Francis, 157, High Street, Elgin, N.B.
Davies, Joseph, 37 & 38, Lister St. Bridge, B'ham.
Davison, Edward, Market Place, Kettering.
Dean, Henry, 71, North Road, Durham.
Dean, W. & Son, 16, Weaman St., B'ham.
Delany, Josph., 8, O'Connell St., Clonmel, Co. Tipperary.
Derrington, Thos. Fredk., 66, Lower Loveday Street, B'ham.
Dickinson, Herbert, 2, Union Row, Minories, E.C.
Dickson, John & Son, 63, Princes St., Edinburgh.
Ding, James, Main St., Ballybay, Co. Monaghan.

Diss, Fred M., 5, Pelham's Lane, Colchester.
Dixon, Cornelius, 348, Witton Road, Aston, B'ham.
Dodd, Thomas, 126, Steelhouse Lane, B'ham.
Doignan, John, 35, Upper Ormond Quay, Dublin.
Dougall, James D. & Sons, 23, Gordon St., Glasgow.
Drew, Henry, Market Place, Romsey, Hants.
Duerden & Hey, 6, Bradley Road, Nelson, Lancs.
Dyke, Frank & Co., 5, 6 & 7, St. George's Avenue, E.C.
Eady, John, 23a, Weaman St., B'ham.
Easson, James, 42, Castle St., Dundee.
Eaton & Co. Lim., High St., Market Harborough, Le'stersh.
Ebrall Brothers, 4, Wyle Cop, Shrewsbury.
Edwards & Son, 2, George St., Plymouth.
Edwards, Benjamin jun., 50½, Newton Street, B'ham.
Edwards, Geo., 25, Lower Castle St., Tralee, Co. Kerry.
Edwards, John, 27, Lower Castle St., Tralee, Co. Kerry.
Edwards, W.H. & Co. Lim.,32 & 33, Weaman St., B'ham.
Edwards, Walter & Co., 4, Whittall St., B'ham.
Ellis, Miss Mary, 22, Darkgate St., Aberystwyth, S. Wales.
Emslie, Samuel, 20, Barrack St., Dundee.
Erskine & Sons, 61, Victoria St., Newton Stewart, Wigtownshire.
Espir, Fernand, 3, East India Avenue, E.C.
Evans, William, 44, Tullow St., Carlow.
Evans, William, 63, Pall Mall, S.W.
Evans, William, Lion St., Brecon, S. Wales.
Ewen, James Walker, 20, Carmelite St., Aberdeen.
Farmer, Richard, 12, North St., Leighton Buzzard, Beds.
Field, Alfred & Co., 35, Cock Lane, Snow Hill, E.C.
Field, Alfred & Co., 77, Edmund St., B'ham.
Fitchew, Arthur T., 75, High St., Ramsgate.
Flemming, M., Arran St., Ballina, Co. Mayo.
Fletcher, F. S., 158, Westgate, Gloucester.
Ford, William, 15, St. Mary's Row, B'ham.
Forrest & Sons (dlrs.), 35, Square, Kelso, Roxburghsh.
Forrest, G. & Sons, 7, Abbey Place, Jedburgh, Roxburghsh.
Fox, Isaac, 4, Upper Bridge St., Canterbury.
Francis, Chas., 9, Long Causeway, Peterboro', Northants.
Franks & Clarke, 153, Villa St., B'ham.
Fraser, Daniel & Co., 4, Leith St. Ter., Edinburgh.
Fry, John, 14, Sadlergate, Derby.
Furlong, Francis Robert, King St., Saffron Walden, Essex.
Gale, Mrs Annie, 20, Joy St., Barnstable, Devon.
Gallyon & Sons, 66, Bridge St., Cambridge.
Gane, Joseph, Cattle Market, Bridgwater.
Garden, William, 122½, Union St., Aberdeen.
Geering, Lewis, 12, Malling St., Lewes.
George, William Joseph, 181, Snargate St., Dover.
Gibbs, Geo., 39, Corn St., Bristol.
Gilronan, Hugh, Arvagh, Cavan.
Goff, Edwin, Dodford, Bromsgrove, Worcs.
Golden, Charles, 7, Northgate, Bradford.
Golden, Wm., 6, 8, 10 & 12, Cross Church St., Huddersfield.
Gooch, Charles, 3, The Wash, Hertford.
Gooch, George, 78, St. Peter's St., St. Albans.
Gow, John R. & Sons, 12, Union St., Dundee.
Graham, George P., 32, Station St., Cockermouth, Cumb.
Graham, John & Co., 27, Union St., Inverness.
Grant, Stephen & Sons, 67a, St. James' St., S.W.
Gray, D. & Co., 36, Union St., Inverness.
Gray, R.M. & Co.,6, Station St., Walthamstow, N.E.
Green, Edwinson Charles & Son, 87, High St., Cheltenham, & 16, Northgate St., Gloucester.

Greener, William W., 22, St. Mary's Row, & 61 & 62, Loveday St., B'ham.
Greener, William Wellington, 68, Haymarket, S.W.
Gregson, James, 59, Penny St., Blackburn.
Griffith, W.B., 25, Ferry Quay St., L'derry.
Griffiths, Charles S., Belmont Bridge, Skipton.
Griffiths, William, 15 & 16, Weaman St., B'ham.
Griffiths, William, 8, Withy Grove, Manchester.
Hall, Christopher, Market Place, Knaresborough, Yorks.
Hall, Frederick, 13, Court, Price St., B'ham.
Hammond Brothers, 40, Jewry St., Winchester.
Hancock, Walter Thomas & Co., 4, Pall Mall Place, S.W.
Hanson, Mrs. Elizabeth, Cornhill, Lincoln.
Hardy Brothers, London & North British Works, Alnwick, Northumb.
Hardy, John Charles, High St., Holbeach, Lincs.
Harkness, William J., George St., Templemore, Co. Tipp'ry.
Harkom, Joseph & Son, 30, George St., Edinburgh.
Harper, Albert J., 127, Steelhouse Lane, B'ham.
Harrison, Thomas, 8, Bank St. Carlisle.
Harvey, James, 16 (back of), St Mary's Row, B'ham.
Hawkes, Thos., Harpers Buildings, Weaman St., B'ham.
Hawkins, Jacob, South St., Ilkeston, Derbysh.
Hellis, Charles, 119, Edgware Road, W.
Hemsley, R., Market Hill, Southam, Warwickshire.
Henderson, James, 32, Barrack St., Dundee.
Henderson, Thomas, Highland Club Buildings, Inverness.
Henry, Alexander Limited, 18, Frederick St., Edinburgh.
Hepplestone, Thomas, 25, Shudehill, Manchester.
Higham, Edward & George, 9, Ranelagh St., Liverpool.
Higham, George G., 20, Berriew St., Welshpool, N. Wales.
Higham, George Garnett, 3, Bailey St., Oswestry, Salop.
Hilliard, Thomas, Bank St., Templemore, Co. Tipp'ry.
Hinton, George, 5, Fore St., Taunton.
Hobson, John, 63, Regent St., Leamington.
Hockey, George Henry, 12, Market Place, Brigg, Lincs.
Hodges, Edwin Charles, 8, Florence St., Isington, N.
Hodges, Lionel, 18, Charterhouse Buildings, E.C.
Hodgett, Joseph, 13, Court, Price Street, B'ham.
Hodgson, Francis, Market Place, Bridlington, Yorks.
Hodgson, Francis, Sewerby-cum-Marton, Hull.
Hodgson, Henry, Hatton Court, Ipswich.
Hodgson, Jesse Parker, 27, Mercer Row, Louth.
Hodgson, William, 8, Middle St., Ripon, Yorks.
Holland & Holland Limited, 98, New Bond St., S.W.
Holland, Thomas, High St., Burford, Oxon.
Hollis, Isaac & Sons, Lench St., B'ham.
Holloway & Co., Vesey St., St. Mary's, B'ham.
Holmes, Henry Charles, 15, Bath St., B'ham.
Homer, Henry B. (finisher), Vesey St., St. Mary's, B'ham.
Hooke, James Alexander, 38 & 39, Pavement, York.
Hooton & Jones, 60, Dale St., Liverpool.
Hooton, William Mouel, South Gate, Sleaford, Lincs.
Horsley, T. & Son, 10, Coney St., York.
Horton, William, 98, Buchanan St., Glasgow.
Hughes, J. & Sons, 18, Bath Street, B'ham.
Hughes, Robert & Sons, 100, Moland St., B'ham.
Hume, George, 6, Loreburn St., Dumfries, N.B.
Hunter & Son, 62, Royal Avenue, Belfast.
Hutchings, James, 9, Bridge St., Aberystwyth, S. Wales.
Hutchinson, Charles & Co., 43, Finkle St., Kendal.
Ingram, Charles, 18b, Renfield St., Glasgow.
Jackson, Samuel, 7 & 9, Church Gate, Low Pavement, N'ham.

James, Enos & Co., 38, Staniforth St., B'ham.
Jeffery, (W.J.) & Co., 60, Queen Victoria St., E.C. & 13, King St., St James, S.W.
Jeffery, Charles, High East St., Dorchester.
Jeffery, Richard, 15, Borough, Farnham, Surrey.
Jeffery, Samuel Richard, 137, High St., Guildford.
Jeffery, William & Son, 12, George St., Plymouth.
Jeffries, Lincoln, 121, Steelhouse Lane, B'ham.
Jewson, Alfred J., 7, Westgate, Halifax.
Johnson & Reid, 29, Post House Wynd, Darlington.
Johnson, Thomas & Son, Market Place, Swaffham, Norfolk.
Jones, Charles, Mariner's Square, Haverfordwest, S. Wales.
Jones, Horatio, 25, High St., Wrexham, N. Wales.
Jones, Robert, 42, Manchester St., Liverpool.
Jones, William Charles H., New Road, Newtown, N. Wales.
Jones, William P., 25, Whittall St., B'ham.
Kavanagh, William & Son, 12, Dane St., Dublin.
Kennedy, John, Jail St., Ennis, Co. Clare.
Kerr, Charles, 74, Hanover St., Stranraer, Wigtownshire.
King, William & Son, 74, King St. West, & 12a, Bridge St., Hammersmith, W.
Knight, Peter, 22, Carrington St., Nottingham.
Knox, H.A. & Co. Limited, 44, St. Mary Axe, E.C.
Lancaster, Charles, 151, New Bond St., W.
Landell, William, 106 & 108, Trongate & Silvan Works, Broad St., Mile End, Glasgow.
Lane Brothers, 45a, New Church St, Bermondsey, S.E.
Lang, Joseph & Son, 102, New Bond St., W.
Langley, James John, 31, Park Square, Luton & 5, Bucklersbury, Hitchin, Herts.
Lawson, James, 70, Argyle St., Glasgow.
Leader Cycle Co., 29, High St., Kingston, Surrey.
Lee, William, 28, Lancaster St., B'ham.
Leech, William & Son, Conduit St., Chelmsford.
Leeson, W.R., 38, Bank St., Ashford, Kent.
Leonard, Daniel & Sons, 133, Steelhouse Lane, B'ham.
Lewis, Charles E., High St., Alford, Lincs.
Lewis, Geo. Ed., 32 & 33, Lower Loveday St., B'ham.
Lightwood, J. Birks, 50, Weaman St., B'ham.
Line Throwing Gun Co. (The), 4, India Buildings, Dundee.
Lingard, Ebenezer & Co., 144, Victoria St. Sth., Gt. Grimsby.
Linington, James, 107a, St. James' Square, & 24, Union St., Newport, Isle of Wight.
Linscott, Tom, John St., Exeter.
Linsley Brothers, Lands Lane, Leeds, & 53, Tyrrel St., Bradford.
Little & Son, 14, Silver St., Yeovil, Somerset.
Liversidge, Chas. Fredk., 29, Market St., Gainsboro', Lincs.
Lloyd & Son, 22, Station St., Lewes.
Lloyd, Henry John, Davygate, York.
London Armoury Co. Lim., 114, Queen Victoria St., E.C.
London Small Arms Co. Limited, Old Ford Road, Bow, E.
Loughlin, Michael, North Main St., Youghal, Waterford.
Loveday, Alfred, East Harling, Thetford, Norfolk.
Lowbridge, Phillip, 16, New Bldgs., Price St., B'ham.
Lynch, J.J., 4, Duke St., Drogheda, Co. Meath.
M'Kenna, Joseph, 8, Essex Quay, Dublin.
MacLeod, John, Tarbert, Loch Fyne, Argyllshire.
Macnaughton, James, 26, Hanover St., Edinburgh.
Maleham, Charles Henry, 5a, West Bar, Sheffield.
Malloch, Peter D., 26, Scott St., Perth.
Marks, James, 61, Queen St., Portsea.
Marks, Joseph, 86, High St., Winchester.

APPENDIX TWO

Marson, John, 14, Weaman St., B'ham.
Marson, Samuel & Co., Great Western Gun Works, Livery St., B'ham.
Martin, Alexander, 20 & 22, Royal Exchange Square, Glasgow, & 128, Union St., Aberdeen.
Martin, John, 24, Peter St., Waterford.
Mason & Beddall, 35, Whittall St., B'ham.
Materface, Henry John, High St., Honiton, Devon.
McCall, W. & Co., 23, Castle St., Dumfries, N.B.
McCarthy, Buck & Co., 11 & 12, St. Andrew's Hill, E.C.
McCririck, James & Sons, 72, Sandgate St., Ayr, & Bank St., Kilmarnock, Ayrshire.
McLagan, Peter, 33, County Place, Perth.
McLoughlin, Charles & Sons, 89, High St., Cheltenham.
McNaughton, James, 44, George St., Perth.
McPherson, Duncan (dealer), 7, Drummond St., Inverness.
Meeham, Thomas, Arvagh, Cavan.
Mellor, Rupert, Upper Tean, Stoke, Staffs.
Metcalfe, Bartholomew, Market Place, Richmond, Yorks.
Midland Gun Co., 80 & 81, Bath St., B'ham.
Millichamp, Charles (field clock guns), High St., Presteign, S. Wales.
Mills, Mrs. Frances, 62, North Lane, Canterbury.
Mirfin, Thomas, 1, Central Chambers, High St., Sheffield.
Mitchell, John & Son, 5, Midsteeple Buildings, High St., Dumfries, N.B.
Monk, William H., 77, Foregate St., Chester.
Montrieux, Theodore, 5, Whittal St., B'ham.
Moody, Charles, Church St., Romsey, Hants.
Moore & Grey, 11, The Arcade, Aldershot.
Moore, (William) & Grey, 165, Piccadilly, W.
Moorhouse, John, 2, Anboro' St., Scarborough.
Morris Aiming Tube & Ammunition Co. Limited (The), 11, Haymarket, S.W.
Morris, Edmund, Cattle Market, Bridgwater.
Morris, Philip & Son, 9 & 10, Widemarsh St., & 4, High St., Hereford.
Morris, William, 19, High St., Enniskillen, Co. Fermanagh.
Morrow & Co., 4, Horton St., Halifax, & 60, Station Parade, Harrogate, Yorks.
Mortimer & Son, 86, George St., Edinburgh.
Morton, W. & Son, 17, Patrick St., Lim'k.
Morton, W. & Son, 53, George St., Cork.
Mountstephen, John Henry, Fleet St., Torquay, & 1, Radford Place, Plymouth.
Murray, David, 23a, St. David St., Brechin, Forfarshire.
Murray, T.W. & Co., 87, St. Patrick St., Cork.
Naylor, Clement, 29, Bridge St., & 6, Woodhead Road, Sheffield.
Needham, Joseph Vernon, Loveday St., B'ham.
Nelson, Francis, 42, Castle St., Sligo.
Nelson, Frank Horatio, Broad St., Chipping Sodbury, Glos.
Newnham, George, 29, Commercial Road, Landport.
Nichols, Alma (dealer), Stalham, Norwich.
Norman, Benj., Church St., Framlingham, Suffolk.
Osborne, Charles & Co. Lim., 2, Great Scotland Yard, S.W.
Osborne, Charles & Co. Limited, 12, 13 & 14, Whittall St. & Sand St., B'ham.
Pagewood, Albert Thomas, 17, Nicholas St., Bristol.
Palmer, Augustus, 31, Preston St., Faversham.
Palmer, Edward, 35, High St., Strood, Rochester.
Palmer, George, 29, High St., Sittingbourne, Kent.
Palmer, W. & H.E., 85, High St., Rochester.
Pape, William R., 29, Collingwood St., & 36, Westgate Road, Newcastle-on-Tyne.
Parker, Alfred G. & Co., 264, Icknield St., B'ham.
Parkes, Frederick, 22, Weaman St., B'ham.
Parkinson, John, 17, Arran Quay, Dublin.
Parkinson, Tom, Market Place, Ulveston, Lancs.
Patstone, John & Son, 25, High St., Southampton.
Percy, Joseph, 48, King St. West, Manchester.
Perkes, Thomas, 96, High St., Eton (Windsor), Bucks.
Phillipson, William, 78, Weaman St., B'ham.
Pierpoint, Ed. J., 1, New Buildings Price St., B'ham.
Playfair, Charles & Co., 142, Union St., Aberdeen.
Pollard, Herbert Edward, 62, Broad St., Worcester.
Pollard, William Hebdon, 63, King William St., E.C.
Powell, Peter, 77, High St., Tonbridge, Kent.
Powell, William & Son., 35, Carr's Lane, B'ham.
Pryse, Thomas, 7½, Bath St., B'ham.
Pulvermann, Martin & Co., 26, Minories, E.
Purdey, James & Sons, Audley House, St. Audley St., W.
Queen Cycle Co. (The), 6, Queen's Road, Battersea, S.W.
Radcliffe, Kenneth Dudley, 150, High St., Colchester.
Ray, T. & Co., 44, Lowfield St., Dartford.
Reilly, E.M. & Co., 277, Oxford St., W.
Rhodes, Frank, 5, North St., Scarborough.
Richards (Westley) & Co. Limited, 178, New Bond St., W.
Richards, (Westley) & Co. Lim., 12, Corporation St., B'ham.
Richards, William, 27, Oldhall St., Liverpool.
Richards, William, 44, Fishergate, Preston.
Richardson, George Benjamin Bank, Barnard Castle, Durham.
Rigby, John & Co., 24, Suffolk St., Dublin.
Rigby, John & Co., 72, St. James' St., S.W.
Roberts, Edgar, 5, Steelhouse Lane, B'ham.
Roberts, Joseph T. & E., 141, Steelhouse Lane, B'ham.
Roberts, Joseph T., 26 (back of), Weaman St. & 5, Steelhouse Lane, B'ham.
Robertson, Alexander & Son, Bridge St., Wick, Caithness-sh.
Robertson, Hugh, Station Road, Dunfermline, Fifeshire.
Robertson, John, 4, Dansey Yard, Wardour St., W.
Robinson, Robert, 7, Queen St., Hull.
Rogers, James Franklin, Market Hill, Woodbridge, Suffolk.
Rogerson, John & Co. Lim., Wolsingham via Darlington & Custom House Chambers, Quayside, Newcastle-on-Tyne.
Roper, Robert, Son & Co., Haymarket, Sheffield.
Rosson, Charles, 4, Market Head, Derby.
Rowe, James, 62, High St., Barnstaple, Devon.
Royal Small Arms Factory, (Col. F.W.J. Barker, R.A., supt.), Sparkbrook, B'ham.
Royal Small Arms Factory, Enfield Lock, Enfield, Middlesex.
Rudd, Arthur James, 54, North St., Norwich.
Russell, Alexander John, 32, High St., Maidstone.
Ryland, Joseph, No. 1 Court, Weaman St., B'ham.
Sanders, Alfred, 79, Bank St., Maidstone.
Saunders, John, 26a, Loveday St., B'ham.
Schmidt, von Max, Alliance Steam Mills, Chapel Road, Stamford Hill, N.
Schwarte & Hammer, 6, Lime St., E.C.
Scotcher & Son, 4, The Traverse, Bury, Suffolk.
Scott, W. & C. & Son, 78, Shaftesbury Avenue, W.
Scott, W. & Co., Lancaster St., B'ham.
Scott, Walter, 19, Marton Road, Middlesborough.
Sell, John Tucker, Potter St, Bishop's Stortford, Herts.
Sellars & Co., 81, High St., Elgin, N.B.
Shaul, William, 3, King St., Tower Hill, E.

Silver, S.W. & Co. & Benjamin Edgington Limited, Sun Court, 67, Cornhill, E.C.
Simonon, Noel, 16½, Loveday St., B'ham.
Slingsby Brothers, 10, High St., Boston, Lincs.
Slingsby, David Powell, 44, Lowerhead Road, Leeds.
Small, Peter, 28, Pilgrim St., Newcastle-on-Tyne.
Smallwood, Samuel, 4 & 5, Mardol, Shrewsbury.
Smith, Alfred & Son, 28, Whittall St., B'ham.
Smith, C.H. & Co., 123, Steelhouse Lane, B'ham.
Smith, Charles & Sons, 47, Market Place, Newark.
Smith, Charles, 25, Weaman Street, B'ham.
Smith, Frederick J. & Co., 54, Clement St., B'ham.
Smith, Midgley & Co., 25, Sunbridge Road, Bradford, Yorks.
Smith, Samuel & Son, Gun Barrel Works, Witton Road, Aston Park, B'ham.
Smith, Samuel, Mil St., Cavan.
Smithson, George James, 6, Scot Lane, Doncaster.
Smyth, John, Water Lane, Donegal.
Smythe, Joseph F., 12, Horsemarket, Darlington; 21, Dovecote St, Stockton-on-Tees & 16, Albert Road, Middlesbro.
Snowie, Hugh & Son, 36, Church St., Inverness.
Snowie, William M., 50, Church St., Inverness.
South of England Gun Co., Harbour St., Folkestone.
Spencer, Charles, Finkle St., Richmond, Yorks.
Stacey & Co., 17 & 19, Settles St., E.
Stensby, Thomas, 20, Hanging Ditch, Manchester.
Stewart, Robert, 2, Magdalene St., Drogheda, Co. Meath.
Stoakes, John, 27, St. Peter's St., Canterbury.
Stoakes, K. & Co., 8, George St., Hastings.
Stovin, William, 4, Westgate, Grantham.
Styles, J.G. & Co., 62, Branston St., B'ham.
Tannahill, Andrew, 1, Smith Hill, Paisley, N.B.
Tarrant, Elijah, 16, Sussex St., Cambridge.
Taylor, Richard, 13, Court, Price St., Lancaster St., B'ham.
Thacker, Thos. (pistol), 28, New Buildings, Price St., B'ham.
Thompson, Herbert, 22, West St., Boston, Lincs.
Thompson, James, Back Row, Hexham, Northumb.
Tilney, Robert & Son, Smallgate St., Beccles, Suffolk.
Tims, Frederick Hope, Cathedral Lane, Truro.
Tisdall, John, 8, South St., Chichester, & High St., Arundel, Sussex.
Tisdall, W.H., 18, Sand St., B'ham.
Tolley, Jas. & Wm., Pioneer Works, Loveday St., B'ham.
Townsend & Williams, 11 & 12, Sand St., B'ham.
Tranter Bros., 18, Sand St., B'ham.
Trevitt, Job, 19, Bridge St., Boston, Lincs.
Troughton, Stephen, 60, Elizabeth St., Blackpool.
Trulock & Harriss, 9, Dawson St., Dublin.
Trulock Bros., 13, Parliament St., Dublin.
Turner, Henry Arthur, High St., Marlborough, Wilts.
Turner, Thomas & J.S. Limited, Fisher St., B'ham.
Turner, Thomas & Sons, 8, Butter Market, Reading, & 86, Northbrook St., Newbury, Berks.
Tuthill, Joseph, North Main St., Youghal, Waterford.
Veals, S. & Son, 3, Tower Hill, Bristol.
Venables, John & Son, 99, St. Aldate's St., Oxford.
Vickers, Sons & Maxim Limited, Erith; Crayford & Dartford, Kent.
Wade, Wallace, Red Lion St., Aylsham, Norfolk.
Wakefield, W.H. & Co., 10, Whittall St., B'ham.
Wales, Durrant, 16, Regent St., Yarmouth, Norfolk.
Walker & Crawford, 72½, Weaman St., B'ham.

Walker, Frank E., 11, Cheap St., Newbury, Berks.
Wallas, William, 66, King St., Wigton, Cumb.
Wallis Brothers, 156, High St., & 4, Corporation St., Lincoln.
Wallis, Richard John & Co., 3, Waterloo St., S.E.
Walton & Co., 25, Leeds Road, Nelson, Lancs.
Wanless, William, 20, Norfolk St., Sunderland.
Ward & Sons, 2, St. Mary's Row, B'ham.
Ward, H.A., Alpha Gun Works, 27, Loveday St., B'ham.
Ward, Peter, Henfield, Sussex.
Warrick, John & Co., 34, St. Mary's Butts, Reading.
Warrilow, James Bakewell, Factory Lane, Chippenham, Wilts.
Watkings & Co., 75, High St., Banbury, Oxon.
Watson & Stewart, 138, Murraygate, Dundee.
Watson Brothers, 29, Old Bond St., W.
Watson, Donald, 19, Inglis St., Inverness.
Watson, Pearson Thomas (executrix of), 97, Clayton St., Newcastle-on-Tyne.
Watson, Rowland, Victoria Gun Works, Whittall St., B'ham.
Webley & Scott Revolver & Arms Co. Limited (The), Shaftesbury Avenue, W.
Webley & Scott Revolver & Arms Co. Limited, 81 to 91, Weaman St., B'ham.
Weekes, Charles & Co., 26 & 27, Essex Quay, Dublin.
Welsh, Samuel, Fermanagh St., Clones, Co. Monaghan.
West, Mrs. Eliza, 11, Grove St., Retford, Notts.
Westley Richards & Co. Limited, Bournbrook, & 12, Corporation St., B'ham.
Weston, C. & H., High St., Hailsham, Sussex.
Weston, Charles & Herbert, 7, New Road, Brighton.
Wheatley, George, Hitchin St., Biggleswade, Beds.
Wheelock, John, Main St. North, Wexford.
Whistler, Edward & Co., 11, Strand, W.C.
Whitby, Henry Albert, 61, Weaman St., B'ham.
White & Popple Limited, Lockhurst Lane, Coventry.
Whitehouse, Edward & Co., High St., Melton Mowbray, Le'stersh.
Whittaker & Co. (Robt. Nazor prop.), 6, Thomas St., Limerick.
Wiggett, John, 74, Bath St., B'ham.
Wild, Thomas & Co., 19, Whittall St., B'ham.
Wilkes & Sons, 26, Manchester Road, Bradford.
Wilkes, John, 1, Lower James St., Golden Square, W.
Wilkes, Joseph & Sons, 26, Woodhouse Lane, Leeds.
Wilkinson Sword Co. Lim. (The), 27, Pall Mall, S.W.
Willcocks, J. Wm., 14, St. Mary St., & 2, Ironmonger St., Stamford.
Williams & Powell, 27, South Castle St., Liverpool.
Williams, Charles D. & Co., 86, Ann St., Belfast.
Williams, Frederick Wm., 32 & 33, Weaman St., B'ham.
Williams, James & Co., 1, Great Hampton St., B'ham.
Williamson & Son, 34, Bull Ring, Ludlow, Salop.
Williamson & Son, 72, High St., Bridgnorth, Salop.
Williamson, Chas. Wm., 7. Bridge St., Stockton-on-Tees.
Willis, George, 1, New Buildings, Price St., B'ham.
Wilson, Edwin, 13, Rampant Horse St., Norwich.
Wilson, George Hy., 9, Market Place, Horncastle, Lincs.
Wilson, James, 47, Goodramgate, York.
Woffindin, John, Market Place, Market Rasen, Lincs.
Wood, James Laidlaw, 33, St. Mary's St., Stamford.
Woodward, James & Sons, 64, St. James' St., S.W.
Wright & Currey, 1, Churchgate, Spalding, Lincs.
Wright, Job, 13, Court, Price St., B'ham.
Yarnold, Wm (finisher), 31½, Whittall St., B'ham.

Bibliography

Books

Instructions to Young Sportsmen in All that Relates to Guns and Shooting, Lt-Col P. Hawker, 1844 (9th ed.)
The Mantons, Gunmakers, W. Keith Neal & D. Back, 1967
Great British Gunmakers, W. Keith Neal & D. Back, 1984
Boss & Co., Builders of Best Guns Only, D. Dallas, 1995
London Gunmakers, N. Brown, 1998
Shooting, Field & Covert, Lord Walsingham & Sir R. Payne-Gallwey, 1887 (2nd ed.)
Something about Guns & Shooting, Purple Heather, 1893 (3rd ed.)
Hit and Miss, Lord Walsingham, 1927
Modern Sporting Gunnery, H. Sharp, 1906
The Causes of Decay in a British Industry, Artifex & Opifex, 1907
Shooting on a Small Income, C. Walker, 1900
Damascus Barrels, J. Puraye, 1968
Yesterday's Shopping: The Army & Navy Stores Catalogue, 1907, David & Charles Reprints, 1969
The Royal Small Arms Factory at Enfield Lock, D. Pam, 1998
California Gunsmiths 1846–1900, L.P. Shelton, 1977
The British Shotgun, Vols: I & II, I.M. Crudgington & D.J. Baker, 1979 & 1981
The Modern Shotgun, Vols: I, II & III, Major Sir G. Burrard, 1947 (2nd ed.)
Great Guns, Lesser Columbus, 1896
How to Buy a Gun, B. Tozer and H.A. Bryden, 1903

Catalogues

Eley's Ammunition, 1899
Kynoch Ltd, 1902–3
C.G. Bonehill, Undated (*c.*1895)
C. Lancaster, 1893
Ward & Sons, Undated (*c.*1905)
W.J. Jeffery & Co., 1904–5
W.R. Pape, Undated (*c.* 1892)

Periodicals

The Field
The Kynoch Journal
The Shooting Times
The Badminton Magazine
Arms & Explosives
The Sporting Goods Review
Land & Water
Country Sport
The Country Gentleman
Bells Life in London

Index

Page entries in italics denote illustrations.
.410 Shotgun 115
Abbotts Waterproofs 89
accident notice 63
Adams, W.A. 20
Afrindia 42
Agnew, A. 68
Albert Edward, Prince of Wales, 12, 74
Albert, H.R.H. Prince 9, 12
Altro 90, *92*, 93
Andrews Ltd, C.W. 42
Anomaly 42
Arms & Explosives 96
Army & Navy Stores, The 26, *29*, 37
Atkin, H. *19*

B.S.A. 113, 118
Badminton Magazine 107
Baker Gun Company 60
Baker, Sir Samuel 42
Bales, G. *11*
Ballistite *98*, 99
Beesley, F. *19*, 22, *109*, *110*, 113
Beresford, Hon. R. 77
Bland, T. *48*, *49*, *50*, 51, 85, 86
Boer War 52
Bogardus Ball 78
Bogardus, Adam H. 78
Bonehill, C.G. 37, 38, *39*, *43*, *45*
Borland, W.D. 99
Boss & Co. *17*, 72
Boss, T. 20, 34
Boswell, C. 23, 26, *27*, *33*, 75, 76, *77*, *88*
Brain, T. 31
Brazier, 24
Browning 108, 109
Brummitt, W. *63*, *64*
Bryden, H. A. 51
Burgess W. *46*
Burrard, Major Sir Gerald 17

Carr Brothers *76*
cartridges 104

Chamberlin & Smith *46*
Champion of Kent, The 120
Chasspot, A.A. *55*, 57
chilled shot 101
Chilton 24
Churchill, E. 76
Clabrough, J. P. & Brothers 30
Clairville 57
Clay Bird Shooting Association *79*
Climax loaders *103*
Climax, The 102
Cody, Col 78
Cogswell & Harrison 37, *78*, 111, *112*, 120
Colindian 42
Colonial quality 37
Cosmos 42
Coton, W.T. *74*
crow starver 62

Damascus barrels 66, 72
de Visme-Shaw, L.H. 107, 119
Demon Works 38
Derby Shot Tower *100*
Dixon, J. 102
Dominion 37

E.C. Powder Co. 99
Ebrall, S. 12
Eley Brothers *95*
Enfield 67
Erskine Loader *102*
Erskine, J. 102
Euoplia 42
Evans, W. *65*
Exhibition, Great 10

Facile Princeps 32, *84*
Feltine 99
Field, The 91, 98, 113
Finsbury Park 73
flight shooting *83*
Ford, W. *117*
Fosbery, Lt-Col G. 42

Gentlemen, The 100 73
George, King, Vth 64
George, W.J. 120
Gevelot Company 92
Gibbs, G. *22*
Gilbertson & Page *54*
Golden Jubilee, 1887 78
Grant, S. 23
Great Depression 118
Greener, W.W. *31*, 32, 34, *84*, 85, 86, *95*
Grey, W. 20
Guest, C. *64*
Gun Club 77, 79, 80
Gunmakers' Arms 106, 114
Gunmaker's Lament, The 96

Hall Ltd, J. & Son *24*
Harrington & Richardson *58*, 59, 117
Hawker, Col Peter 18, 20, 77, 105
Heath, C. J. 92
Heathcote, Frank 73, 74
Hill & Smith *106*
Hill, A. & Son *28*
Hodges, E. 23
Holland & Holland 15, 37, *41, 114*
Holloway & Co. *26*
Hornsey Wood House 73
Horsley, T. *27*, 52, *67, 116*
Houllier, M. 95
How to Buy a Gun 51, 67
Hurlingham Club 74, 77, 80

Inanimate Bird Shooting Association 79
Interchangeable 38, *39*, 41, 117

Jeffery, W.J. & Co. 26, *40*, 41
Johnson, I. 59
Joyce & Co, F. *97*
Jungle 42, 43

Keeper's Gun 49, 51
Keeper's Gun, Hammerless *50*, 51
Kench, The *87*
Kennedy, Lord 73
Kingdom-Murrill, Dr W. *87*
Kynoch Journal, The 99
Kynoch 64, 89, 91, 96, 105

Lancaster, C. *20*, 22, *35*, 37, *103*, 108, 113
Lancaster, C.W. 42
Land & Water 80, 81
Lang, J. *16, 30*, 113
Liege 53, 55, 59

Ligowsky, G. 78
live pigeon trap shooting 74
locks *70*

Manton, J. *10,* 18, 20, 21, 42
Match, Anglo-American 79
Midland Gun Co., The 38
Milbank, F. 65
Monstrum Horrendum 35
Monte Carlo 77
Moore, W. 20
Morris Tube Company 43, *44*, 45
Murcott, T. *13*

Newcastle Chilled Shot Company 101
Nobel, A. 24, *29, 77*, 98
Nobel's Ammunition Factory *101*
Normal Powder Company 89

Osbaldeston, Squire 73
Osborne, C. 59

Pape, W. R. *17, 28, 47, 85*
Paradox *41*, 42
Patent Act, 1852 10
Payne-Gallwey, Sir Ralph 35
Perfect, The 89, 105
Perkes, T. 70
Phillips, H. 113
Pioneer Works, 85
Premier Works 31
Prentice's gun cotton 98
Pulvermann & Co., Martin *61*
Punt gun *87*, 88
Purdey, J. *18*, 22, *47, 82*
Purple Heather 20
Pygmie 93, *105*

Rigby & Bissell patent 90
Rigby J. *90*, 108
Robertson, J. *32*, 34
Rogers, J. 29, 30
Royal Flying Corps 118

Sandhurst 37
Schultze *77*, 98, 99
Scott, W. & C. 29
Shikari 42
Shooting on a Small Income 25, 111
Shooting Times, The 54
Smith, G. *51*
Snider 67
Spencer 108

Sporting Goods Review, The 106
Sporting Magazine 73
Squires, J. *69*
Stanton 24
Starkie's Chemist 23
Stevens, J., Arms & Tool Co. *59*
Swiftsure *78*

Tabatière *56*, 57
Target gun 118
Thorn, Alan *113*
Thorn, H.A.A. 37, 78, *103*, 113
Times, The 99
Tolley, J. & W. *84*, 85, 86, *90*, *92*, *93*, 120
Tordus 53
Tozer, B. 51, 52
Twentieth Century Gun, The 118

Ubique 42

Vaughan, C.B. 35, 37
Vena Contracta 111
Vesdre 53
Victoria Small Arms Co., The *38*
Victoria, Queen 78

Walker, C.E. 25, 111
Walsingham, Lord 72
Ward 118
Warrilow, J.B. *94*
Webley, T. 29, 31, 113
Westley Richards *30, 31*
Whistler 37
Wild West Show 78
Wilkes, J. *66*
Winchester Model '97 109

York, Duke of 64, 72

Zulu, The *56*